Green Framing

An Advanced Framing

How-To Guide

Green Framing
An Advanced Framing How-To Guide
First edition, 2010

By Tim Garrison, P.E., The Builder's Engineer™

Feedback to the author is encouraged via the Forum at www.constructioncalc.com. Or via U.S. mail: ConstructionCalc, Inc., 18579 West Lakeview Lane, Mount Vernon, WA 98274.

ISBN 1450585965

Photographs in the Introduction by author's family. All other photographs by the author. All illustrations and solved examples by the author.

This book is dedicated to thinkers

It doesn't take much thought
to do what's always been done. To build the same old way
because you've been doing it that-a-way for 50 years.

It takes courage
to learn. To try different things - more efficient methods. To
those who think that way, this book is for you.

CONTENTS

INTRODUCTION

I'm not one for fads or the latest fashion but even I have to admit build-green is a hot topic. For me, however, it's always been hot. I'm a cheapskate, and building green saves money, plain and simple. I've been a green proponent my entire career and, boy, am I glad the rest of the world has finally caught up to me.

A quick word about my background. I grew up on a 76-acre cattle ranch. My three brothers and I learned all manner of construction there. We remodeled the old house then built a new one; built steel pipe corrals; repaired the barns; built bicycles, miles of fences, dune buggies, '65 Mustangs; you name it.

Building my parents' home in 1979.

What's with those shorts?

Couple that with bachelor and master degrees in civil engineering and lots of years pounding nails as a carpenter. Then add nine years as a staff engineer in the public sector, then seventeen as a business owner of construction and engineering companies. Top it all off with starting and still running a software company, ConstructionCalc, Inc.

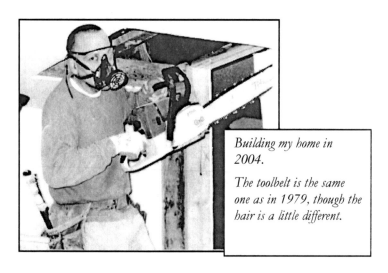

Building my home in 2004.

The toolbelt is the same one as in 1979, though the hair is a little different.

Mix all of the above into a big stew kettle and you wind up with the perfect recipe for this book. I hope you enjoy reading it as much as I did writing it.

As you work though these pages you'll undoubtedly have comments or questions. I'd love to hear them. The best way to interact with me is through my forum at www.constructioncalc.com. It's free and educational.

Disclaimer

There is a lot of engineering advice in this book. Bear in mind, it is just that: advice. It's not intended to work in every situation, so use good judgment when applying.

The author takes no responsibility or liability for problems, disasters, failures, etc., which may arise from the application of principles or concepts presented herein.

DEFINITIONS

The construction industry is rife with jargon and technical terms. To make sense of this book, you'll need to understand a few key words and phrases.

A stickler may note that the following definitions are not exactly what you'd find in Webster's. That is because they're intended to be working definitions—what you really need to know when talking turkey with a building professional.

Admixture: Something added to concrete during batching (mixing together of the ingredients) other than Portland cement, water, or aggregate. Admixtures enhance certain properties of normal concrete, such as: high early strength (*accelerators*), resistance to water intrusion (*waterproofers*), resistance to freeze-thaw (*air entrainment*), and flowability (*plasticizers*).

Bearing capacity: A material's strength relative to its ability to support something. For example, soil must have adequate bearing capacity to support a footing or the footing will settle. As another example, the top of a post must have adequate bearing capacity to support the beam sitting on it, or the post top may crush.

Bottom chord: The bottom member of a truss, usually horizontal (scissors trusses, however, have sloped bottom chords).

Cantilever: A structural member supported at one end, with a free, usually overhanging, end at the other. For example, a beam with an overhanging end; or a floor joist with one end extending beyond its bearing wall; or a post embedded in the ground with its other end not connected to any lateral-resisting element; or a retaining wall fixed rigidly to its footing and with no connection to a floor at its top.

Collar tie: A horizontal framing member, usually 2x4 or larger, connected to opposing roof rafters. Collar ties are almost never recommended as structural members, but may be used effectively to hold a ceiling for an attic space.

Concrete, cement, etc.: Concrete is comprised of a powder—called Portland cement—aggregate (sand and/or gravel), and water. Leave out the aggregate and you have "neat cement." As wet concrete cures, it undergoes a chemical reaction called hydration. It is improper to say you have a cement driveway or cement basement. Make that faux pas and everyone will know you are a rookie. Once Portland cement is mixed with water and aggregate it is *always* called concrete.

Continuously braced: When a compression member (column, post, stud) is so well braced over its entire length that it cannot buckle sideways.

Also when the compression side of a bending member (beam, joist, rafter) is continuously braced, as by sheathing, such that it cannot buckle sideways. See *unbraced length*.

Cripple: Short vertical members, usually 2x material, used as wall framing where full-length studs would be too long. Examples include, over a header, under a window sill, as short "studs" in a pony wall (a.k.a. cripple wall). When over a header, cripples transfer gravity load from the top plate above to the header. When under a window sill cripples are non-structural, acting as backing for drywall or siding.

Dead load: A gravity load due to the weight of permanent building materials. Examples include roofing, flooring, studs, joists, and trusses. Structural design treats live and dead loads differently, that is why they are broken out separately.

Deflection: The distance a structural member or assembly moves when load is applied. For example, a floor joist may deflect, or sag, 0.5 inch under full load. Deflection of structural members is usually expressed in terms of L/some number, where L is the member's length in inches. For example, most beams are limited by code to a deflection, under live + dead loads, of L/240. Say you have a 15-foot-long beam. Code would limit its maximum deflection under full load to $15*12/240 = 0.75$ inch.

Diaphragm: See horizontal diaphragm.

Differential settlement: Settlement that is of different amounts over a structure's footprint. All structures built on soil settle a little, but as long as that settlement is uniform over the footprint, usually there are no problems. But if one portion of a building settles significantly more than the rest, this differential settlement is cause for concern.

Dormer: A small sub-roof system penetrating through a main roof system. The chief purpose of a dormer is to provide a window to an attic or upper floor through one of the main roof planes.

Factor of safety: A measure of extra strength. For example, things built in compliance with most codes have a factor of safety of 2.5, meaning they are 150 percent stronger than necessary to avoid breaking.

Flange: The top or bottom portion of an I-joist or wide flange beam; or the top or bottom portion of any structural member with web-flange configuration (the web being connected to the flange(s).

Gable roof: A type of roof-wall system where an end wall extends vertically past ceiling height, into the shape of a triangle (in elevation view), terminating at a roof. This is different than a hip roof, which has no gables, but rather the roof terminates at ceiling level all the way around its perimeter.

Glu-lam: Short for glued-laminated. A structural member, beam or column, typically, which is constructed of multiple plies of dimension lumber, normally 2x stock, glued together.

Gravity Load: Any load that acts downward due to weight. Examples are snow, live, and dead loads. See also lateral loads.

Gusset: A structural covering plate, usually plywood in wood-framed construction, or steel plate in structural steel construction, that connects the ends of two or more structural members.

Header: A beam over a door or window.

Heel: The place in a truss where top and bottom chords come together. Also the low end of a rafter where it bears on a beam or wall.

Hip: A portion of roof, other than at a ridge line, where two intersecting planes of the roof come together. Similar to a valley, except a hip has an angle between the planes of joined roof greater than 180 degrees, whereas the angle between valley planes is less than 180 degrees.

A **hip roof system** consists of hips and valleys, no gables, with its eaves generally at the same elevation—typically at ceiling level.

Horizontal diaphragm: A floor or roof, usually, that collects lateral forces (wind, seismic, or earth) then transmits those forces to shear walls or other lateral force-resisting elements. Ceilings can be horizontal diaphragms, too, if so designed and detailed. The connections between horizontal diaphragms and lateral force-resisting elements must be at least as strong as the structural elements themselves.

Hydrostatic pressure: The pressure exerted on a surface by something liquid. This is the same pressure your eardrums feel at the bottom of a swimming pool. As another example, groundwater exerts lateral hydrostatic pressure on basement walls and upwards hydrostatic pressure on basement floors. Also, wet concrete in a form exerts outward hydrostatic pressure on said forms. Hydrostatic pressure increases with increased fluid depth.

ICF: Insulated Concrete Forms. A building product resembling a large, hollow expanded polystyrene Lego Block, used to construct walls. These are fitted with rebar in their hollow cores, then poured full of concrete. The polystyrene outer shells stay in place, providing insulation and nailing surfaces for drywall and siding.

I-joist: An engineered wood product comprised of wood flanges top and bottom, joined with a plywood or OSB web in between. These are commonly used as floor joists or roof rafters. TJI is a common brand by Trus Joist MacMillan™.

In plane: In the same plane as the wall. See *out of plane* for more.

Joist: A horizontal framing member used to construct a ceiling or floor. Joists are normally spaced 16 inches, 19.2 inches, or 24 inches apart.

King stud: A full-length stud abutting the end of a beam or header. King studs are generally nailed to trimmers. Trimmers provide vertical support to the ends of the beam or header. If no trimmer is used, the king stud must provide gravity support to the header or beam.

King studs must resist out-of-plane wind loading tributary to the adjacent window or door.

Lateral load: Force applied in a sideways direction, usually from wind or earthquake. Soils can also apply lateral loads to retaining walls. *Gravity* loads are downward acting due to gravity acting on a system's mass.

Lift: In earthwork, a lift is a thin layer of soil, usually no more than 12 inches thick. When soil is used to fill a low spot, the proper way is to fill in lifts, compacting each lift as you go.

Live load: A gravity load from the weight of non-building materials. Examples include cars, snow, people, furniture, and workmen. Structural design treats live and dead loads differently, and that is why they are broken out separately.

Load path: One of the most important concepts in structural design. Applied loads, either gravity or lateral, must be resisted from their point of application all the way to the ground. This path, through various structural elements, is the load path. The connections joining the various structural elements must be as good as the elements themselves or a weak link in the load path chain exists.

LSL: Laminated Strand Lumber, made from wood chips epoxied under high pressure. Similar to PSL except different, weaker wood chips and or glues are used.

LVL: Laminated Veneer Lumber, made from thin layers of wood sheets bonded with epoxy under high pressure. They look like a plywood beam and have similar strength to a PSL.

Moment: Having to do with bending. For example, if you snap a pencil in half, you have just applied a bending moment which exceeded the pencil's allowable bending stress. Mathematically, a moment equals a force times a distance. Beams, joists, and rafters typically resist moments (as well as shear forces). As another example, if a connection of a beam to column can transfer bending forces, i.e. the connection is not pinned, it is **moment-resisting**.

In structural members such as beams, joists, and rafters, an internal moment (within the member) consists of tension on one side and compression on the other. For example, a simply supported floor joist has compression in the top flange and tension in the bottom flange. The compression side must be braced against sideways movement (buckling) or it can easily become unstable and flop over.

Moment Frame: Columns and beams connected together with fixed, moment-resisting connections intended to resist in-plane lateral loads. In order for a frame to be truly moment-resisting its connection(s) must be "fixed", i.e. designed to transfer moment from member to member (as opposed to a "pinned" connection, which can not transfer moment). The location of the moment-resisting connection(s) can be at any column to beam, and / or at any column to footing.

A *portal frame* is a moment frame around a door or other opening.

Negative Moment: In a beam, joist, or rafter, negative moment results in compression in the *bottom* of the member and tension in the *top*. This occurs at the supports of cantilevered members and members continuous (no joint) over a support. Negative moments are dangerous because it is easy to forget lateral bracing where the compression occurs, with the result being the member can kick (buckle) sideways and come off its support or become unstable.

Simply supported beams, joists, and rafters, on the other hand, do not have negative moments, i.e. the compression portion is in the *top* of the member and tension is in the *bottom*. Usually there is something attached to the top (joists, decking, or sheathing) that keeps the compression side from buckling.

O.C.: On Center. Usually refers to the spacing, from center to center, of repetitive structural members such as joists, studs, or trusses.

OSB: Oriented Strand Board. Similar to plywood but made from wood chips and glue. Commonly used for roof sheathing, wall sheathing, subflooring, and the web portion of I-joists.

Out of plane: Usually refers to lateral load acting perpendicular to the wall. This is the type of load in a wind storm, for example, that tends to blow walls inward or suck them outward. Conversely, *in plane* usually refers to lateral forces from wind or earthquake acting in the same plane as the wall. In-plane lateral forces cause racking and are resisted by shear walls, moment frames, and other lateral force resisting elements.

Pitch: The slope of a roof or other surface. Pitch is expressed in terms of something and 12. For example a common roof pitch is 4 and 12, or 4:12. This means four units of rise per twelve units of run. This equates to 18.26 degrees; 12 and 12 is equal to 45 degrees.

Pony wall: A short wall usually to fill a space not applicable for a full-height wall. Also called a *cripple wall*. An example is the short section of wall between foundation and main floor, as the foundation steps down a slope.

Portal frame: A moment-resisting frame around an opening, usually a door. Primarily resists in-plane lateral loads but must also resist out-of-plane lateral loads on the door.

Psf, plf: Pounds per Square Foot or Pounds per Lineal Foot. Both are units of load applied to structural members. For example, snow load where I live is 35 psf. A typical residential floor live load is 40 psf. The load used to design a floor joist is the psf loading multiplied by the joist's spacing, say, 40 psf x 2 feet = 80 plf. A typical wind load is 25 psf, laterally.

PSL: Parallel Strand Lumber, made from wood chips epoxied under high pressure. Much stronger than regular, sawn wood. Typically used for beams.

Rafter: A sloped roof framing member, usually one of many in repetition. Rafters must be vertically well-supported at each end. If not, the member is either a truss top chord or is structurally unstable. "Well-supported" typically means the rafter bears on a substantial beam and/or a bearing wall. A *collar tie* does not provide support to a rafter.

Reentrant corner: A corner protruding into a building; an interior corner.

Ridge beam: A horizontal roof beam along a ridge line. With a true rafter system, the ridge beam must be a strong structural element. If the ridge beam is but a slender board, then each rafter will have outward thrust at its low end, which must be resisted by a strong connection to a ceiling joist or other laterally-restraining element. A true ridge beam rafter system has no outward thrust at the rafter's low end.

Shear: 1) In structural design, shear is used as a noun, that is, a stress within a member tending to cause diagonal cracking, particularly near supports. This type of shear has two components: vertical and horizontal-due to bending. The vertical component would tend to cause a beam, for example, to shear off, i.e. drop vertically if it broke at its support. Horizontal-due to bending shear can be thought of this way: if you make a beam out of a bunch of planks stacked on top of each other but don't nail or glue them together, the beam won't take much weight. Under load, the planks sag, sliding relative to each other. Now, however, if you nail or glue the planks together such that they act as one, you can place much more weight on this composite beam. The nails or glue are providing horizontal shear resistance within the composite beam.

2) When discussing wind and earthquake loads, shear is the primary in-plane stress in a shear resisting element such as a shear wall or moment frame. This type of shear causes racking.

Shear wall: A wall that resists in-plane lateral forces (racking forces) brought on by wind or earthquake (seismic) loads. Shear walls can be made of many materials: plywood, OSB (Oriented Strand Board), gypsum drywall, diagonal planking, concrete, and masonry, to name the most common. Of course, each material has different shear strength. For a shear wall to function, it must be well-connected to a horizontal diaphragm and or foundation at top and bottom.

Simply supported: Usually refers to a beam or joist that bears at each of its ends and is not continuous over any supports. In other words, said beam or joist has no cantilever, nor does it have any interior support(s). This makes a significant difference during structural analysis (see "Negative moment").

Sister(ed): Nailing one board over top of another or side-by-side with another such that the two work together to resist load. Can be thought of as an additional ply.

Slab on grade: A concrete slab placed on, and supported by, the ground. This is different than an *elevated slab* which is supported by beams and columns.

Stinger: Slang for a type of vibrator used to consolidate (compact) wet concrete.

Subfloor: The first layer of sheathing, usually plywood or Oriented Strand Board (OSB), attached to the top of floor joists. Carpet, tile, or other flooring is usually placed over the subfloor. The subfloor normally acts as a horizontal diaphragm, transferring lateral loads to shear walls or foundations below.

Subgrade: Native soil below a structural element, usually supporting said element. For example, subgrade below a road is the native soil underneath the road.

TJI: Trade name for an I-joist manufactured by Trus Joist MacMillan.

Top chord: The top-most member(s) in a truss.

Tributary width: The width of attached framing from which load is applied. In other words, the "loaded width." Tributary width is generally measured perpendicular to the member being designed, and for normal stick-framed construction is half the distance from the member being designed to the next supporting member (load is shared equally between supporting members, half goes to one support, half to the other.)

Load is commonly brought to a member from two sides. For example, the tributary width on a floor beam is half the span of the joists on one side plus half the span of the joists on the other.

The tributary width on a typical joist or rafter is those members' on-center spacing.

Tributary area is similar except the load being brought comes from *two* perpendicular directions rather than one and results in a point load in pounds rather than a line load in pounds per foot.

Trimmer: Usually 2x material, similar to a stud but not full-length, that supports the end of a header or beam. Trimmers are typically nailed to king studs. Trimmers provide little resistance to out-of-plane wind loading.

Unbraced length: 1) In design of compression members (studs, posts, and columns) this is the length between points of lateral support. It is over a column's unbraced length that buckling can occur. The shorter the unbraced length, the more compressive load a member can take.

2) Unbraced length also pertains to the compression side of a bending member. For example, a simply-supported joist has compression in its top and tension toward the bottom. If no sheathing is nailed to the top of the joist, it can easily buckle sideways. The sheathing in this case effectively shortens the joist's "compression flange" to what is called *continuously braced*.

Web: 1) An interior member of a truss; i.e., not the bottom chord or top chord, but one of the members—vertically or diagonally oriented—joining the two. In wooden trusses, webs normally have only tension or compression forces, not bending or shear.

2) The middle part of an I-joist, wide flange, or other flanged structural member, joining the top and bottom flanges.

Web stiffener: An additional piece of structural material, usually the same type and thickness as the web, connected directly to the web of a flanged structural member. The purpose is to enhance the web's resistance to buckling or crushing. Typically only necessary under heavy loading conditions, and typically only installed in the vicinity of supports and at point load application areas.

Wide flange: A type of steel member that looks like an "I" in cross section—sometimes generically called an I-beam. Wide flanges are the most common structural steel type.

Chapter 1
Why Haven't These Buildings Toppled?

Introduction

Green framing, advanced framing, optimum value engineering (OVE), frugal framing, call it what you will. In the end it's all about building smart - saving money and resources.

Billions of dollars and vast natural resources are wasted every year in over-built structures. Excessive wood, concrete, and steel are consumed because a faulty "more-is-better" mentality has evolved - a mentality borne of misinformation and perpetuated by fear.

This book illustrates, in plain language, powerful green framing techniques. Roofs, walls, and floors, with their subparts and pieces, are carefully examined. Wasteful practices are exposed, efficient ones suggested. You'll learn how to actually design rafters, beams, studs, and footings using a computer.

The benefits of green framing include:

- Saving thousands of dollars in materials and manpower on every project.

- Conserving significant natural resources - wood, concrete, and steel - and the energy consumed producing them.

- Using less wood leaves more room for insulation. Significant energy savings result over the life of the structure.

- When you cut out unnecessary wood, concrete, and steel, earthquake forces on a building diminish proportionally.

With this knowledge any builder can break outdated wasteful habits, and instead use the right materials in the right places. Any designer can confidently produce efficient, code-compliant plans.

A Little Background

Building codes are a relatively new invention. The first widely-accepted building code in the U.S. was written in the early 1900s. Today's building code, the International Building Code (IBC), has its roots in the Uniform Building Code (UBC) which was first published in 1927. Over the years many local jurisdictions adopted building codes but many did not. Even today there are jurisdictions in the U.S. that do not issue building permits nor require adherence to any building code.

Where I live in western Washington, building codes are strictly enforced for any structure from a shed to a fence to a sky scraper. Around here it's unthinkable that a house might be designed by a non-professional and built without a building permit.

But where my brother lives in Kansas, there are no such requirements. Draw up your plans on a napkin, grab your hammer and go. There are lots of places like that in our country today.

So in America we've got quite a mish-mash of structures. A few that meet current codes but many, many that don't.

I took some photographs of old buildings in my county.

The first one, I call the Titanic. This house is at least 50 years old and as you can see has settled terribly. The house is likely built partially over an old slough that was filled with logs and other debris. The part built over the slough embankments has not settled but the part built over the fill-debris has. This is called differential settlement. Incredibly, people still live in this house.

The next structure is a 75+ year-old commercial building. By today's standards it contains not a single shear wall nor a horizontal diaphragm. It is listing about a foot out of plumb, yet there it stands.

Here is a very large barn, probably 50+ years old. Note how huge its wind sail area (roof) is. Also you can see that it is located in the middle of an open valley with no trees or other buildings to shield it from wind gusts. The gable end walls are mostly door openings, and the wood panels in between don't come close to any sort of legal shear wall. The roof isn't a legitimate diaphragm. There's a two-foot sag in the roof at the eaves. Yet year after year, winter after winter, storm after storm, this barn continues to serve.

According to its historic placard, this building was constructed in 1890. It has undergone an extensive tenant improvement, but other than new windows and doors, the exterior walls, floor and roof framing are original. It is built partially over a salt water channel, supported on timber piers. The horizontal siding on the long walls shows settlement up to a foot in several areas. The above photo is the rear wall. Note all the windows and doors (read: no shear panels.)

The front wall is pretty much the same, all windows; which counts for nothing in resisting lateral (wind and earthquake) loads. Here is what this wall looks like from the inside.

Same wall, about mid-height:

This wall is constructed of horizontal siding attached to 2x4 studs. Not one shear panel, holdown, or hurricane clip.

Roof framing is 2x6 rafters, originally spanning 20+ feet. There is no ridge beam. I'd go so far as saying there isn't one code-compliant piece of lumber or connection in this entire building. And in fact most structural elements are overstressed, according to current code, by several hundred percent.

In its 119-year life, why hasn't this building imploded or blown over?

This last building was also built in 1890. One corner (the one by the streetlight) has settled at least six-inches. But that's not what makes this one of the most dangerous buildings in the county. The front wall is all glass. No shear walls, no portal frames, no buttress walls, nothing. And the next parallel interior wall is some 30-feet back into the building. As the one corner sinks, the building tilts causing racking (shear) stress in the window wall. Should a window break or crack there is a real possibility that this building would fall over sideways – I've seen it happen to a building of similar construction in a nearby town. Yet, this building stands.

All of the aforementioned structures have lived through snow accumulation of several feet, howling wind storms, and earthquakes.

All across America and the world are buildings that don't come close to meeting current code. It usually takes a hurricane, tornado, severe neglect, freakish snow storm, or 7+ magnitude earthquake to bring them down. And even then many survive.

So what's the point?

The point is that many stick-framed structures not built to code are strong enough, and those that are built to code are stronger than they need to be in most cases.

If you live in a jurisdiction that has building codes and enforces them, you don't have a choice but to comply with those codes. *But you don't need to overbuild.*

Let me say that again in a slightly different way. *Our building codes contain so much factor of safety, no one should feel compelled to exceed them.* The grossly non-code-compliant buildings on the previous pages, in my opinion, provide stout testimonial.

Our industry should be actively searching for ways to trim our designs so that they just comply with code and no more. If we build stronger than code we're literally throwing away money and effort. And we're not building green.

This book is about minimal, yet code-compliant, structural design. Green design. The trick is understanding the underlying structural concepts: where loads come from; where they go; and how they're resisted. With that knowledge, we can maximize efficiency and save money.

Chapter 2
Wasteful Practices

The Problem

Not only have I been a framer and been guilty of many wasteful practices myself, I see inefficiency and waste *every time* I walk a jobsite:

- Too-big beams and headers

- Too much blocking

- Too many studs

- Too many trimmers and king studs

- Too many cripples

- Too many holdowns

- Too many shear walls

- Too many posts and piers in crawlspaces

Let's look at some real-life examples. I find it interesting that the following photos are from jobsites I simply walked on, picked at random. I didn't have to search and scour my county for wasteful framing – it's everywhere.

As we go through these, you might think, *Two wasted studs and one oversized header. Big deal.* And you'd be right if those were the only wasteful instances in the house. But it's never just one or two, it's always twenty or more. Add them all up and we're talking significant material waste, big time lost opportunity for insulation, and many hours of manpower down the drain. To the tune of *thousands of dollars*. Billions nationwide annually.

I don't go into much theory in this chapter, that'll come later. At this point we're just identifying typical problems.

The first picture is from the house I studied for the article I wrote for Green Builder Magazine, _Are You Overframing Walls?_, December, 2006. [The original version of this article is entitled, _How Much Overdesign, Really?_, and is available under the Free Items tab at www.constructioncalc.com]

This sliding door opening has a 6x10 header and double trimmers at each end. Single trimmers would have worked, saving two studs. The header could have been a 4x8 which would have used about half as much timber, plus there would have been room left for insulation.

Had the framer lowered the header to the level of the door opening, he could have used less lumber framing it in. Instead of five cripples and two horizontal 2x members, just three cripples, equally spaced, would have done the trick.

Similar situation, same house. The 6x10 header could have been much smaller, a 2x4 would have worked here. Why? Because this wall is at the gable end of the house and the only load this header will ever see is a tiny amount from a gable truss.

What about those eight cripples above the header? Their main purpose is to transmit gravity (downward) load from the gable truss, above, into the header. Two cripples spaced evenly would have been plenty.

There are two trimmers at each end of the header. In this case no trimmers at all were needed. What? Yes. The load in the header is so small that 16-penny nails could transmit the load to the king studs.

How about those five cripples under the window? Their sole purpose is to provide backing for drywall and siding. Two would have been plenty.

The framer of this shear wall used two rows of double, 2x6 blocking. The reason for blocking in the first place was to provide backing for the OSB sheathing. Had the framer oriented one row of sheathing vertically, he could have done away with one row of blocking. Also, this wall is quite long and has tremendous shear capacity. Consequently the blocking could have been single 2x rather than double 2x. Understanding these concepts could have saved a lot of wood and hassle.

This 2nd floor beam is stupendously inefficient. There is a steel flitch plate bolted between two large PSLs, using at least 50, 5/8" all-thread bolts. I would not have wanted to be the guy doing all that drilling and fitting. Presumably, the rationale was to keep the bottom of the beam flush with the rest of the ceiling. Okay. But is there a simpler, more efficient way? Yes – use a steel, wide flange beam. This could have been a W10x22 with a 2x nailer on top and would have fit nicely in the ceiling cavity. It would have not only cost less, it would have actually weighed less and been easier to install.

Another greener option would have been a deeper Glu-Lam beam. It would have extended a few inches below the level of the ceiling, but is that really a problem? It's a matter of personal taste. My house is done that way and I like it (so does my wife, which is the true test.)

The beams break up the expanse of ceiling, adding a little architectural relief to an otherwise big flat surface.

Here we see an example of cripples under a window that are not needed. The distance from window sill to mud sill is less than 24-inches, thus neither the drywall on the inside or the sheathing on the outside need intermediate backing. The possible exception would be if there is a vertical OSB joint under the window which would need backing because this is a shear wall.

Also notice that this window has double sill, double trimmers, and double king studs. Cleaning all this up saves at least six studs – and we didn't even look at the oversized header.

Here we have a corner assembly upon which some framer went ape with his nail gun. All of the corners in this house were built similarly, using around 150 nails per. There are several ways to make a wood-framed corner, some greener than others. This example shows how not to do it. There's too much wood, not enough insulation, and all those nails are doing nothing structurally.

In this single-story commercial building, every glu-lam beam was supported by steel tube columns with heavy, bolted steel column caps. In every case two studs, or three at the most, could have been used instead, with cheap, light-gage metal connectors.

Also we see that the door openings use a double 2x flat header with trimmers each side. A single 2x flat and no trimmers would have worked just as well.

Why space studs at 16-inches? Why not 24-inches and save a bunch of timber?

I'm wondering why even have a glu-lam? Since there is a wall directly beneath, why not make that a bearing wall? This building has slab-on-grade floor, so to make this wall bearing would have only required thickening the slab and adding a little rebar. Of course substantial headers at openings would have been required, but in the end a lot of money, time, and material would have been saved.

This double PSL beam is supported on 8, 2x6 studs. Three studs or a single 4x6 would have worked just fine.

At the other end of this beam, three, 4x6s were used as a composite post. One would have been sufficient.

Over-supporting this beam wasted a small tree's worth of lumber – expensive wood that may just have well been tossed into a landfill.

There sure are a lot of studs here. More than double the number needed. A lot more insulation could have been used if all that unnecessary wood wasn't in the way.

This wall has almost no gravity (downward) load on it. You can tell by the direction of the roof rafters. Knowing that, we also know that the 6x8 window header is way oversized. Also the double trimmers are not needed and neither is the strap at the end of the header.

Gap

This is an interior shear wall. Top, above; bottom, below.

This shear wall is nearly useless because it is not connected to the 2nd floor diaphragm at the top of wall. The wall stops at the level of the bottom of the ceiling joists leaving a gap between it and the element which brings the lateral load, the 2nd floor diaphragm. So all the plywood, nailing, blocking, and holdowns will never see much force and are wasted.

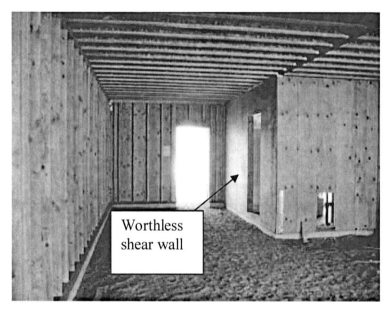

Worthless shear wall

Here is another unnecessary shear wall but for a different reason. See the long exterior wall parallel to this one at the left of the photo? That shear wall has so much capacity that it makes this one unnecessary. So why is it there? Probably because of a prescriptive code requirement. Houses designed per the IRC (International Residential Code) are required to have shear walls throughout, interior and exterior, at certain intervals. That prescriptive requirement does not take into account overdesign like we see here.

How Did We Get Billion-Dollar-Wasteful?

Here's how:

- **Codes**. Building codes have gotten more and more restrictive over the years. They've also become so bloated and difficult to use that most building industry folks avoid them like the dentist. The result is we're gun-shy about efficiency. Today's code mentality is "more is better." So when we aren't sure, we throw in more. Tons and tons and tons more. We may as well throw most of that "more" into a landfill; it does no good at all. And in fact, a lot of the time it is counter-productive: more wood, concrete, and steel means less insulation; and more weight adds proportionally to seismic forces. There is no code-incentive for efficiency.

- **Builders**. Most builders have no training in basic structural theory. It's tough to question a more-is-better mentality when you're not really sure of the underlying concepts. How do builders learn their trade? From other builders; who learned from other builders before them, and so on. Where's the formal structural training? It's never been there.

- **Architects and Designers**. Most architects and designers don't receive enough structural training to make them experts. They generally know enough to size a beam or post but to really sharpen the pencil and get efficient puts them out of their comfort zone. And why go there when the building code doesn't require or encourage it?

- **Engineers**. Having spent the majority of my career supervising other engineers and practicing the trade myself, I know something about how engineers think and behave. Most are more worried about liability than saving someone else's money (the owner's.) They have little incentive to produce efficient designs. To an engineer, more is safer. It takes extra time to explore green alternatives, and with engineers especially, time equals money. Why should an engineer cost himself more money, incur more liability, and go against the grain of the code when he can snow job the owner as to how massively strong he's made the building? Owners don't know to ask the right questions, and the engineer grins all the way to the cruise ship.

- **Building Officials**. Building officials have zero incentive to enforce or even encourage green techniques. They answer to the building code (see first bullet point.)

It's a racket and vicious cycle that desperately needs fixing. The first step is education. Once builders, designers, architects, and code officials understand what's going on and that the solutions are attainable by them, they will start putting pressure on the engineers. Or, better yet, they'll start implementing the designs themselves.

In the BC days (Before Computers) it was unreasonable to expect non-engineers to perform structural calculations. But with the advent of computers and user-friendly software, now anyone can do basic structural design.

A word of caution on this. Most states have laws that require "Engineering be done by licensed engineers," or something similar. Sizing a beam, joist, footing, or post could be defined as engineering. But if that truly were the case, non-engineers wouldn't be able to use span tables. Lumber yard employees wouldn't be able to design floor joists and beams. Architectural designers wouldn't be able to write the sizes of structural members on their plans unless stamped by an engineer. Building officials wouldn't be able to recommend beam sizes. And so on. The fact is, those things happen all the time in every state.

I bring this up just to make you aware that law does exist which describes the practice of engineering and who may practice it; and that if you do your own structural design, you may be required to have it reviewed and stamped by a licensed engineer. Or not...

A Note On Software

Throughout this book we'll use ConstructionCalc™ software for our green designs. You could use any brand. I just happen to be biased toward ConstructionCalc™ because I think it is the most cost-effective, the simplest, and provides the most alternative solutions.

We'll use three programs: ProBeam™, Column, Post, Stud Calculator™, and FootingCalc™. At the time of this book's publication, 2010, each costs $98. At www.constructioncalc.com, there is also a bundle offer: six programs for the cost of two, plus a free pdf copy of the book "Basic Structural Concepts for the Non-Engineer". Shop around and you'll find that's quite a bargain.

Regardless of which brand you buy, you'll be money ahead once you've learned a few basics. A couple hundred dollars spent on software is insignificant compared to the savings realized by framing green.

If you're new to structural design aids (span tables or software) there's a lot of free information at www.constructioncalc.com. In particular check out the blog, forum, solved examples and white papers.

Chapter 3
Roof Framing

A Quick Recap

To summarize what we've learned so far:

1. Building codes set minimum standards of design - standards which include a large factor of safety. We need not exceed those standards.

2. Wasteful practices are everywhere, and occur mainly because building industry professionals do not understand basic structural concepts. There exists a "more is better" mentality, which, with most framing practices, is flawed.

Now let's jump to the meat and potatoes of green framing, the structural design. We'll start with roofs and work our way down.

Trusses

Pre-engineered trusses are among the most efficient, green structural devices. They enable very long spans with a bare minimum of materials. If only the rest of a structure could be so efficient.

Competition among truss manufactures is the catalyst for this efficiency. For example, if XYZ truss Inc. uses 2x6 top chords where 2x4s will work, they won't be in business long. Also, to illustrate Recap Point 1, above, you can be sure that truss companies shave their designs to the bare-bones minimum allowed by code. If they didn't, they wouldn't win many bids. With all those "minimal" trusses nationwide, are there epidemic truss failures? No. About the only time a properly designed and installed pre-engineered truss fails is when a tornado or hurricane rips one apart. And even then, the failure is generally not the truss itself but its connection to the wall.

Design of trusses is complex and well beyond the scope of this book. The thing to keep in mind is that for most roof systems, pre-engineered trusses will be cheaper and greener than stick-framed and thus are preferred.

Rafters

Where trusses won't work, rafters are usually called for. Here are the main criteria that go into the design of a rafter:

- o Span
- o Spacing
- o Loading
- o Depth for insulation purposes
- o Deflection criteria

The strategy for any green design is to determine which variables we can tweak toward the most efficient end. To understand this, let's actually design a rafter.

Rafter Design Example:

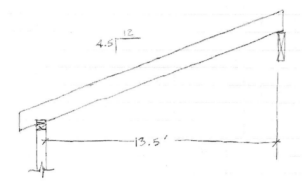

Span. Not much to fiddle with in this case. If there were an interior bearing, it would be a great way to reduce the span and thus rafter size. But seeing none in the sketch, we're stuck with the given span.

Of note is the cantilever (overhang). It's about a foot long. Cantilevers can govern the design so they should always be included.

Spacing. Rafters are typically spaced at 24", a dimension driven by the plywood or OSB sheathing on top and the drywall on the bottom. Here is a green opportunity. What if we used 32-inch spacing? We'd use less wood and more insulation. Let's try it.

First let's check the plywood sheathing. I went to www.apawood.org, Engineered Wood Construction Guide – Roof Construction, Table 29, and found that 32-inch spacing works for APA 32/16, 15/32", or 1/2" sheathing with edge support, for up to 30 psf live load. "Edge support" need not be blocking, it can be panel edge clips. I also checked IBC table 2306.3.1 for horizontal diaphragm and found that there are no restrictions on rafter spacing until 32-inches is exceeded. Our preference, 1/2" plywood roof sheathing, checks.

What about drywall ceiling? I searched all over the web trying to find span tables for gypsum drywall ceilings and found none. It appears that 24-inches is the maximum support-to-support spacing for most national brands of drywall. So if we want a drywall ceiling, our maximum rafter spacing will be 24-inches. However, if we can use something different, such as 1x T&G boards, our 32-inch spacing is okay.

Loading. This rafter will be subjected to a 25 psf snow load (dictated by location and local building department) and 15 psf dead load (composition or metal roofing + gyp ceiling + self-weight of rafters). Some designers like to shave dead load in order to minimize rafter or truss size. Meaning they'll calculate to the tenth of a pound the dead load of all the parts and pieces of the roof + ceiling assembly. I do not recommend that. Principally because I know re-roofing happens and very few roofers strip off the old roofing first. So what might be a 12.8 psf dead load new could well be a 17.2 psf dead load after 30 years. I use 15 for design which easily accounts for one re-roofing. This judgment call is one case where green-design takes a back seat to practicality and safety.

Depth For Insulation. Many times the depth of a rafter is not controlled by strength but instead by insulation requirements. Architects will frequently spec a 12" deep rafter when an 8" or 10" deep member would calc. It might be possible to reduce depth if more efficient insulation material, such as sprayed-in or rigid, were used, but we leave that call to the architect.

Deflection Criteria. From Table 1604.3 in the 2006 IBC, deflection for roof members supporting nonplaster ceilings is limited as follows:

- Snow, Live, or Wind load: $l/240$

- Dead + Live: $l/180$.

These limits will be input in the next step. (The "l" is the rafter's length in inches.)

Calc It. Keep in mind that with ProBeam™ each red triangle conceals a helpful tip or bit of input data that will save you from having to look things up in reference books.

Here's the general input for our rafter at 2-foot spacing:

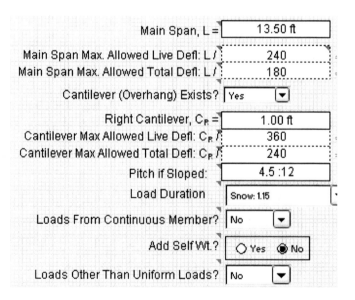

Main Span, L =	13.50 ft
Main Span Max. Allowed Live Defl: L /	240
Main Span Max. Allowed Total Defl: L /	180
Cantilever (Overhang) Exists?	Yes
Right Cantilever, C_R =	1.00 ft
Cantilever Max Allowed Live Defl: C_R /	360
Cantilever Max Allowed Total Defl: C_R /	240
Pitch if Sloped:	4.5 :12
Load Duration	Snow: 1.15
Loads From Continuous Member?	No
Add Self Wt.?	◯ Yes ⦿ No
Loads Other Than Uniform Loads?	No

And here is the loading input. Note that this is a Uniform Load over the Full Length of Member, typical for rafters and joists.

r Full Length of Member		Tributary
	Live, psf	Dead, psf	Width, ft
ıot including snow)		15 psf	2.00 ft
Roof Snow (only)	25 psf		2.00 ft

Also note that our snow load is greater than code-prescribed live load (16 psf) and with ProBeam™ we don't need to input the live load.

Done. Let's check results. First sawn lumber. Note that we took Repetitive Member credit because our joists are spaced 24" or less.

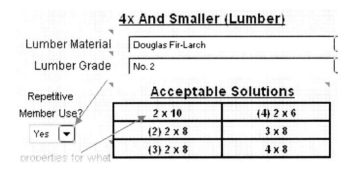

4x And Smaller (Lumber)

Lumber Material | Douglas Fir-Larch

Lumber Grade | No. 2

Repetitive Member Use?

Yes ▼

Acceptable Solutions

2 x 10	(4) 2 x 6
(2) 2 x 8	3 x 8
(3) 2 x 8	4 x 8

properties for what

Here are I-joist solutions.

I-Level, TJI

11-7/8" TJI / L65	9-1/2" TJI 110
16" TJI / L65	9-1/2" TJI 210
16" TJI / L90	9-1/2" TJI 230
14" TJI H90	11-7/8" TJI 360

The logical choices are the 2x10 Doug Fir No. 2, or 9-1/2 TJI 110.

Now let's see what happens if we increase spacing to 32" O.C.

The only two inputs that change are Tributary Width and Repetitive Member credit.

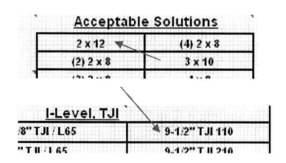

We find that the Doug Fir option jumps to 2x12 but the TJI remains the same.

The question becomes which is the most efficient?

If the architect will not allow anything less than 12-inch nominal depth for insulation purposes, we could use 2x12s or

11-7/8" TJI 110s at 32-inch spacing (9-1/2" TJI 110s work, so by inspection, 11-7/8" will too.) With this option we need APA 32/16, 15/32-inch thick roof sheathing with edge support, and a ceiling other than drywall.

If the architect allows 10-inch nominal rafter depth, we could use 2x10 Doug Fir at 24-inch spacing; or 9-1/2 TJI 110s at 32-inch spacing with the same constraints as above.

One last note, all of the above assumed an *interior* rafter. But this could have been a rafter in a barn or outbuilding, in which case the spacing would not be limited by insulation or by a drywall ceiling. Rafter spacing of 32" O.C., or 40", or even 48" would be possible depending on the type and thickness of roof sheathing. With our 25 psf snow load, APA 40/20, 5/8" thickness with edge support works for 40" rafter spacing, and 48/24, 3/4" thickness with edge support is okay for 48-inch spacing. Calc'ing such an exterior rafter is as simple as changing the Tributary Width to the new spacing, and reducing Dead Load (I'd use 8 psf for the roof system without insulation and drywall.) For wind and earthquake resistance, horizontal (roof) diaphragms are okay with framing members spaced up to 48-inches O.C. as long as field nailing is increased from 12- to 6-inches O.C. (2006 IBC Table 2306.3.1 footnote b).

This example illustrates not only how to design a rafter but also how to tweak certain design variables to the most efficient end. This process works for any rafter anywhere. Just be sure to use the correct loading, span, pitch, etc.

Beams

Stick-framed roof systems typically consist of rafters supported by bearing walls and by beams. Beams can be located at ridges, hips, valleys, and in ceilings. Green design requires sizing beams large enough to meet code but no larger.

Here are the main criteria that go into the design of any beam:

o Span

o Loading

o Depth for insulation or architectural purposes

o Deflection criteria

Let's look at a couple examples, starting with a ridge beam.

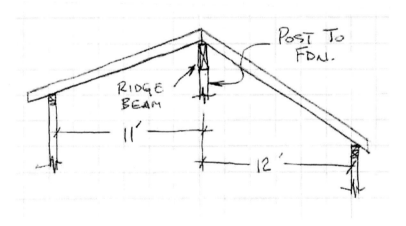

Span. This is about the only variable we can fiddle with in our effort to be green. In general, the shorter the span, the smaller the ridge beam. But, also, the shorter the span, the more support posts and footings we'll use. For this example, let's say our ridge beam must span from gable to gable, the entire length of the house, say 30-feet. At each gable end the beam

---

will be supported on triple studs that extend down to the foundation.

Loading. We'll use the same loads as the rafter example:

Snow: 25 psf

Live: 16 psf, but snow controls.

Dead: 15 psf. This does not include the weight of the ridge beam itself, called self-weight. We must add that, but it's a click in ProBeam™.

There isn't much opportunity to green the design by tweaking loads. They are prescribed by code.

Tributary Width. The loads are applied to the beam from the high ends of the rafters. The tributary width of the rafters on the left is half their span: 11'/2 = 5'. The tributary width of the rafters on the right is half their span: 12'/2 = 6'. The total tributary width on the beam, then, is the sum from each side: 5'+6' = 11'. Again, there isn't much we can do to lessen this unless bearing wall(s) were added under the rafters. But that's not shown, so the ridge beam must carry the rafters. Note that the ridge beam carries half the rafter's total load. The other half is carried by the exterior walls. This is always the case in standard, light-framed construction.

Depth. In our example, we're not constrained by the depth of the beam – it can be as deep as needed. This is very important because such a long span will require a deep beam.

Deflection Criteria. As in the rafter example, deflection is code-limited to l/240 for snow load only and l/180 for total load (snow + dead).

Here is the ProBeam™ General Input. It's very similar to the rafter example - notice that many inputs don't change. With a little practice you'll get familiar with the program and find that designs go very quickly, usually less than a minute. I envision the day when green builders will park a laptop computer at every jobsite.

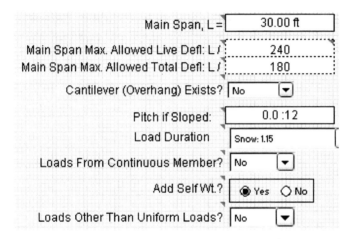

Main Span, L =	30.00 ft
Main Span Max. Allowed Live Defl: L /	240
Main Span Max. Allowed Total Defl: L /	180
Cantilever (Overhang) Exists?	No
Pitch if Sloped:	0.0 :12
Load Duration	Snow: 1.15
Loads From Continuous Member?	No
Add Self Wt.?	● Yes ○ No
Loads Other Than Uniform Loads?	No

And here is the loading.

r Full Length of Member			Tributary
	Live, psf	Dead, psf	Width, ft
ıot including snow)		15 psf	11.00 ft
Roof Snow (only)	25 psf		11.00 ft
Floor 3 Loads			

The smallest Doug Fir sawn beam that works is a 10x24 – too big and expensive so let's ignore that.

Here are the glu-lam choices.

24F-V4 (DF/DF)	▼

Acceptable Solutions

-	5.125" x 18"
3" x 22.5"	6.75" x 16.5"
3.125" x 21"	8.75" x 15"
5" x 18"	

A reasonable option would be the 5x18, though any are acceptable.

What about PSL? Here are our options.

2.0E Parallam PSL

-	5-1/4" x 18"
-	7" x 16"
-	SИ VИ= 35

We see that the sizes are about the same as glu-lam. A potentially significant difference between glu-lam and PSL is that camber (a.k.a. crown) is built in to glu-lams but not necessarily into PSLs. This means that the PSL will have a small sag under its own weight + dead load whereas the glu-lam starts with a slight crown and will flatten out under self-weight + dead load.

Lastly, let's consider steel. ProBeam™ lists tubes and wide flanges. Wide flange options are shown on the next page.

If beam height were an issue, steel would be the way to go. We could use an 8" tall, W8x48 (the 8 is the nominal height in inches and the 48 is the weight in lbs/ft.) This, however, is not nearly as efficient as a W12x22, which is four inches taller but weighs less than half as much.

A992, Fy=50 ksi ▼

-	W14x22	-
-	W16x26	-
-	W18x35	-
W8x48	W21x44	-
W10x30	W24x55	-
W12x22	-	

To summarize, our 30' span beam is quite large regardless of material. Let's see what happens if we add a post in the middle, cutting the span in half, to 15'.

We need only change the span in ProBeam™ - everything else remains the same.

Main Span, L = | 15.00 ft |

Now we have a lot more options:

- o 6x12 Doug Fir No. 1, or
- o 3x10.5 glu-lam, or
- o 2-11/16x11-1/4 PSL, or
- o 1-3/4x14 1.9E LVL

It's interesting that by cutting the span in half (factor of 2), we cut the amount of material by a factor of nearly 3.

A valuable lesson here is that the relationship between span and beam size is not linear, it's geometric. Meaning a small increase in span causes a big increase in beam (or joist or rafter) size. Thus to keep those members small, we'll keep spans short whenever possible.

A second, perhaps more subtle lesson is that the relationship between beam height and strength is not linear, it too is geometric. Meaning a small increase in the height of a beam (or joist or rafter) causes a big increase in that member's bending strength. This was borne out in the steel example: a relatively small increase from 8" to 12" inches in height resulted in a whopping 55% reduction in steel. Thus to minimize materials, use members as tall as possible.

The above two lessons apply to all bending members: beams, rafters, and joists. We'll revisit this throughout the remainder of the book.

Now let's look at another type of roof beam, a hip beam.

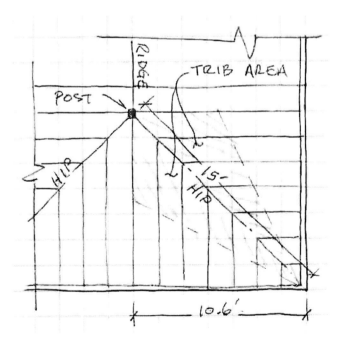

The design criteria are the same as for the ridge beam:

- o Span

- o Loading

- o Depth for insulation or architectural purposes

- o Deflection criteria

Span. We're given a span of 15-feet, from the exterior wall to a support post where the hip intersects a ridge (note this dimension is the horizontal projection, not the true length.) The post can extend to its own footing or to a beam in the ceiling. From our previous example we know that to minimize the size of our hip beam, we want to minimize the span and maximize the height of the beam. We're not given any opportunity for additional mid-span bearing so we'll calc this beam as-is.

Our hip beam slopes at a pitch of 5:12.

Loading. We use the same code-prescribed loads as previously.

Snow: 25 psf

Live: 16 psf, but snow controls.

Dead: 15 psf, not including self-weight.

Tributary Width. The loads are applied to the hip from the high ends of the rafters on each side. In the sketch you can see that the tributary area is triangular-shaped. This is called a "wedge load" and is maximum at the high end of the hip and goes to zero at the low end. In ProBeam™ there is a section specifically for inputting this type of load.

First, the General Input.

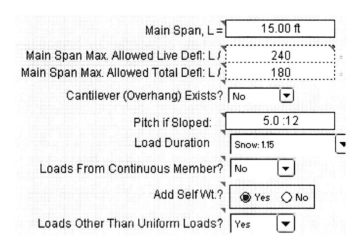

Main Span, L =	15.00 ft
Main Span Max. Allowed Live Defl: L /	240
Main Span Max. Allowed Total Defl: L /	180
Cantilever (Overhang) Exists?	No ▼
Pitch if Sloped:	5.0 :12
Load Duration	Snow: 1.15 ▼
Loads From Continuous Member?	No ▼
Add Self Wt.?	⦿ Yes ○ No
Loads Other Than Uniform Loads?	Yes ▼

Here is the loading. Notice that we used a different section in ProBeam™ for this "Wedge Load".

ɹe Loads On Main Span Only (Max at Left End, Zero at Righ

	Live Load, psf	Dead Load, psf	Tributary width, ft
Wedge Load A	25 psf	15 psf	10.60 ft

This loading section doesn't have separate input cells for snow and live loads so we choose the heavier of the two, snow load, and input it as live.

The tributary width on *each side* of the hip beam is half the span of the longest rafter: 10.6'/2 =5.3'. The rafters are symmetrical on each side of the hip beam so the total tributary width is 5.3' + 5.3' = 10.6'.

Deflection Criteria. As in the ridge beam example, deflection is code-limited to l/240 for snow load only and l/180 for total load (snow + dead).

Here are several Doug Fir solutions.

Lumber Material	Douglas Fir-Larch	
Lumber Grade	No. 2	

Repetitive Member Use?	**Acceptable Solutions**	
	-	(4) 2 x 10
No ▼	(2) 2 x 14	3 x 16
	(3) 2 x 12	4 x 12

t roperties for what

We could also use 6x10 Doug Fir, as well as glu-lam, PSL, LVL, and other options.

Our final selection might depend on what we've got laying around the jobsite or in our boneyard. If we had a bunch of 2x12s, we might use three of them nailed together.

Here's an option that I like especially well:

1.55E Timberstrand LSL	
1-3/4" x 11-1/4"	(3) 1-3/4" x 9-1/4"
(2) 1-3/4" x 9-1/4"	Sif VM= 8.5

This 1-3/4 x 11-1/4 LSL is relatively cheap and probably something we have on hand. It's also a single member as opposed to 3, sawn members nailed together.

Let's select this as our Final Member, like so.

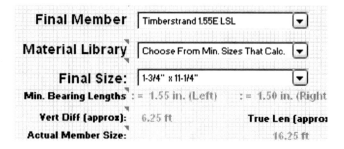

Final Member	Timberstrand 1.55E LSL ▼	
Material Library	Choose From Min. Sizes That Calc. ▼	
Final Size:	1-3/4" x 11-1/4" ▼	
Min. Bearing Lengths	:= 1.55 in. (Left)	:= 1.50 in. (Right)
Vert Diff (approx):	6.25 ft	True Len (appro:
Actual Member Size:		16.25 ft

Now look to the right of the screen and check efficiency.

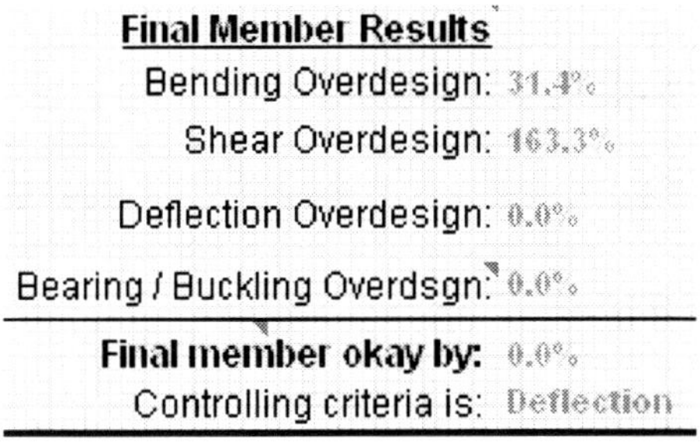

Final Member Results
Bending Overdesign: 31.4%
Shear Overdesign: 163.3%

Deflection Overdesign: 0.0%

Bearing / Buckling Overdsgn: 0.0%

Final member okay by: 0.0%
Controlling criteria is: Deflection

Here we see that this LSL has extra capacity in bending and shear but is maxed out for deflection, with 0% overdesign. *This is exactly what we're looking for in our quest for green design. Members that calc, but just barely.* (Bearing / Buckling only applies to I-Joists, so doesn't apply to this LSL.)

The ability to check efficiency is a powerful feature of this software. With just a couple or few clicks you can check hundreds of alternatives and select the most efficient, or maybe just use the one you have on hand.

Chapter 4
Wall Framing

A Quick Recap

To summarize what we've learned so far:

o *Green design conforms to building code but doesn't necessarily exceed it.*

o *More is usually not better.*

o *Beams, rafters, and joists are at their most efficient when: 1) Span is minimized; and 2) Member height is maximized.*

o *Use software to tell you how efficient a member is. Select options with low percentage of overdesign.*

o *Sometimes the greenest option is the one laying around in your boneyard. Use up that "waste".*

Now let's examine wall framing.

Top Plates

Stick framed walls have employed double top plates as standard practice for decades. But do top plates really need to be doubled? If not, several hundred feet of 2x material could be saved with every home.

Let's see what the building code has to say.

• 2006 IBC (IBC 2308.9.2.1) requires double 2x top plates on all bearing and exterior walls with a few exceptions.

This sketch shows code-compliant, standard double top plate construction.

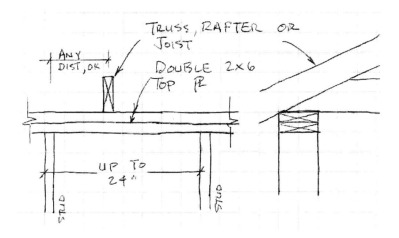

EXCEPTION. *Single top plate can be used when studs line up within 1" of supported truss heels, rafters, or joists; and joints, corners, and intersecting walls are tied with 3"x6" galvanized plate.* (IBC 2308.9.2.1, Exception). Like so.

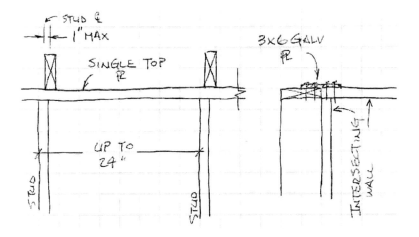

- When using double **2x4** top plates with studs spaced at 24" OC, and supporting joists or trusses are spaced at >16" OC, joists or trusses must be within 5" of studs or use triple top plate. Note this does not apply to **2x6** walls. (IBC 2308.9.2.2)

- Non-bearing walls can use single top plates. (IBC 2308.9.2.3)

In summary, non-bearing, interior walls can use single top plates; and bearing walls can use single top plates IF studs line up with trusses, rafters, joists, AND galvanized steel plates are used to tie corners and intersections together.

But there's more to this story. When wind and seismic forces are large (>350 pounds per lineal foot), shear walls must be constructed with double 2x or single 3x boundary members. Top plates are common shear panel boundaries, thus if the lateral loads are high, double 2x (or 3x, but 3x material can be hard to find and expensive) is required regardless of whether studs line up with joists and or trusses (IBC Table 2306.4.1, footnote i).

Another consideration is the case where interior walls use single top plate but exterior walls use double. To ensure that top plate elevations are constant, different length studs would be required – interior studs being 1-1/2" longer than exterior. I question whether this is a greener option than simply using double top plates everywhere.

Bottom line: In regions of high wind or seismic forces, double top plates are nearly always recommended. In other areas, single top plates may be used, but serious consideration should be given to the difficulty and expense in lining up trusses, joists, and rafters, and using metal tie plates.

Studs

Most stick-framed walls use 2x studs at 16" spacing, which is usually plenty strong. But is it plenty green?

The answer is no, for a couple of reasons. From a structural standpoint, there is nothing sacred about 16" stud spacing. As long as studs are able to resist lateral and gravity loads, they can be spaced farther apart than 16", common alternatives being 19.2" or 24".

Less wood in a wall allows more insulation. The more insulation the better (greener.)

As an example, say we have a single-story house of 2,000 square feet, 50'x40'. The approximate length of exterior wall is 180 lineal feet. If studs are spaced 16", we'll use about 135 of them. If they're spaced at 24", we'll only use about 90. That's a savings of 45 studs - significant. Not to mention we're able to replace nearly six lineal feet of solid wood, floor to ceiling, with insulation.

And of course the same principles apply to interior walls.

The key to knowing a stud's allowable spacing is knowing its structural capacity. Here are the main design criteria:

- o Unbraced length
- o Spacing
- o Gravity loading
- o Lateral loading

o   Deflection criteria

To get a feel for this, let's actually design a typical stud. We'll use "Column, Post, Stud Calculator"™ by ConstructionCalc™.

Unbraced Length. Any compression member; stud, column, or post, is much stronger if short rather than long. To prove it, here's a fun little experiment - the spaghetti test. Take a piece of uncooked spaghetti and orient it vertically on a table. Push down on the high end. With very little effort the spaghetti bows in the middle and with continued force breaks. This phenomena is known as "buckling". Now take one of the small pieces and perform the same test. The short piece can take much, much more force, even though it has exactly the same cross section ("meat").

Buckling is one reason that building codes require engineering analysis for any stud over ten feet long. Another reason has to do with wind load, but we'll get to that later. *The point is short studs can carry more load than long ones of the same cross-section.*

What about the word "unbraced" in Unbraced Length? If we go back to our original piece of long spaghetti and have a helper grip it between finger and thumb about mid-length, restraining it from moving sideways, we find that quite a bit of force is required to cause a bow. Maybe not as much as the short piece but definitely more than the unbraced long piece. Bracing the spaghetti against lateral movement effectively shortens it, which, in turn, strengthens it. The unbraced length is the longest distance between points of lateral bracing.

Now, concerning our example stud. Let's say it is 8' long, one of many studs in a wall. It is connected to drywall on one side and sheathing on the other. Those two items provide lateral bracing in the stud's weak axis. In other words, the drywall and sheathing keep the stud from buckling in the direction parallel to the wall. But is there anything to keep the stud from buckling in its strong axis? In other words can the stud bow inward or outward, perpendicular to the wall? Yes it can, there is nothing to stop it from buckling in that direction.

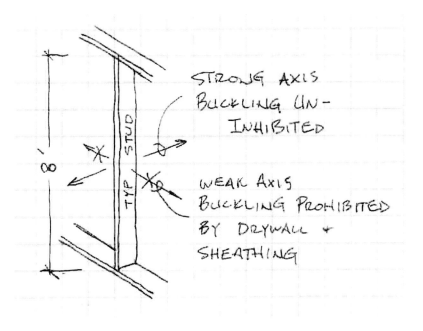

Now let's get to our input.

formation

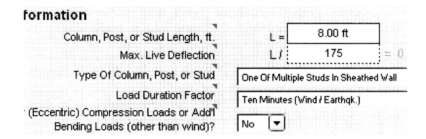

Column, Post, or Stud Length, ft.	L = **8.00 ft**
Max. Live Deflection	L / **175** = 0.
Type Of Column, Post, or Stud	One Of Multiple Studs In Sheathed Wall
Load Duration Factor	Ten Minutes (Wind / Earthqk.)
(Eccentric) Compression Loads or Add'l Bending Loads (other than wind)?	No ▼

The length, 8', is the unbraced length.

Max Live Deflection. This is the amount we're willing to allow the stud to bow. L/175 equals the length in inches divided by 175 and is suitable to ensure that plaster or any other attached hard cladding or window will not crack under full load.

Type of Column, Post, or Stud. In our case we're designing a typical stud so we select that. This indicates to the program that weak axis buckling is restrained but strong axis buckling is not.

Load Duration Factor. Our stud is on an exterior wall, exposed to wind. We select the load factor for the shortest duration, worst case loading, Wind / Earthquake.

Off-Center, Eccentric Load (other than Wind). If our gravity loading (downward load) is applied to the top of our stud, which it is, we select No. If the load was applied off-center, such as by a ledger, we'd select Yes.

Now loads.

First, *Gravity Load.* It's not shown in the sketch but let's say that this wall supports roof trusses spanning 40' with a 1' eaves overhang. Here is the gravity load input.

	Live, psf	Dead, psf	x Length, ft	x Width, ft.
ds (without snow)		15 psf	1.33 ft	21.00 ft
Roof Snow (only)	25 psf		1.33 ft	21.00 ft

We're using the same snow and dead loads as in the previous chapter.

The tributary area is the loaded area that one stud supports. The "x Length" is the stud's spacing, 16" (equals 1.33 ft.). The "x Width" is half the truss' span + overhang.

Here is the calculated gravity load supported by a single stud.

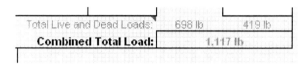

Total Live and Dead Loads:	698 lb	419 lb
Combined Total Load:	1,117 lb	

Now let's input wind load.

A little background. Calculating wind pressure per the IBC is quite complicated, beyond the scope of this book. But we can make reasonable estimates as follows.

- o Light wind load = 20 psf

- o Medium wind load = 30 psf

- o Heavy wind load = 40 psf or more.

The IBC recognizes that maximum wind load will not likely occur at the same time as maximum snow load, so allows half of one while the other is 100%. A proper analysis requires that both scenarios: half wind + full snow, and half snow + full wind, are examined.

We'll check full snow + half wind first. We've already input full snow load, now let's assume our full wind load is 30 psf. Half of that equals15 psf.

Here's our input.

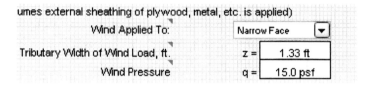

Wind Applied To. Wind will blow against the narrow face of our stud, so we select that.

Tributary Width of Wind Load. This is half the distance from our stud to the nearest stud on one side + half the distance to the nearest stud on the other side: 16"/2 + 16"/2 = 16" or 1.33'. Note that this equals the stud's spacing, which is always the case for evenly spaced studs.

Wind Pressure. As discussed above, in this run we'll use half the total wind pressure, 30 psf/2 = 15 psf.

Done. Now let's check results.

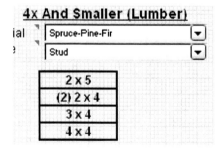

I've chosen an inexpensive material and grade, Spruce-Pine-Fir (SPF), Stud grade, and we see that a 2x5 would work. By inspection so would a 2x6, which is what we originally intended.

Now to hone the design.

Let's see what happens if we use 24" spacing rather than 16". We'll make only two changes. Gravity Load "x Length" and Wind Load tributary width will both go to 2.

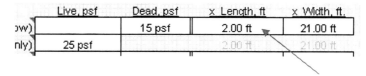

	Live, psf	Dead, psf	x Length, ft	x Width, ft.
ɔw)		15 psf	2.00 ft	21.00 ft
nly)	25 psf		2.00 ft	21.00 ft

And…

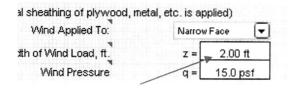

al sheathing of plywood, metal, etc. is applied)

Wind Applied To:	Narrow Face ▼
ɟth of Wind Load, ft.	z = 2.00 ft
Wind Pressure	q = 15.0 psf

Checking the allowable solutions, we find that a 2x6 SFP, Stud grade works.

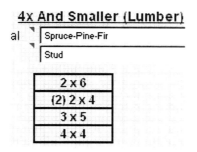

4x And Smaller (Lumber)

al | Spruce-Pine-Fir

Stud

2 x 6
(2) 2 x 4
3 x 5
4 x 4

And checking further, we find that our 2x6 SFP, Stud grade makes it by quite a comfortable margin, 42.9%.

This member makes it by: **42.9%**

Controlling Factor: **Combined Bending / Compressive Stresses**

The only thing left to do is check the load combination, 1/2 snow + full wind. We make two input changes: reducing the snow load to 12.5 and increasing the wind load to 30.

	Live, psf	Dead, psf	x Length, ft	x Width, ft.
v)		15 psf	2.00 ft	21.00 ft
y)	13 psf		2.00 ft	21.00 ft

This program automatically rounded 12.5 up to 13. Even though 13 is showing in the input cell, the computer *is* using the input value, 12.5, in computations.

And…

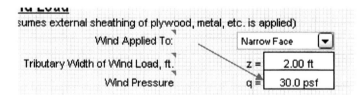

Our 2x6 SPF, Stud grade still makes it, but now only by 15.4%. Still, we're code-compliant and okay.

So to summarize, for our example we can use inexpensive stud material at 24" spacing. Very green.

Now let's see what happens if we raise the ceiling height to 10'. The only input change is unbraced length.

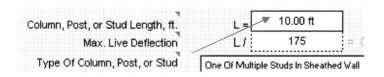

Not surprisingly we find that our stud size jumps to 2x8.

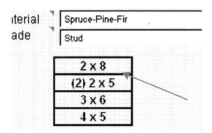

No good. Let's try a better grade of SPF. I tried No. 3 and it didn't make it but No 2 / 1 did.

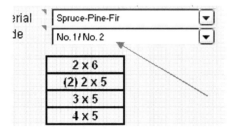

If we wanted to keep with Stud grade we would have had to reduce the spacing back to 16", or maybe 19.2". Try it.

As you can see, the number of variables is great, so great in fact that without a computer, zeroing in on the greenest solution would be impractical. With a computer, however, and good software, it's quick and easy.

Corners and Intersections

A lot of wood gets burned up in the construction of corners and intersections. There are several ways of building these, some greener than others.

The design variables include:

- o   Gravity load capacity
- o   Shear wall nailing
- o   Space for insulation

When I was a framer in the 70s and 80s, this is how we constructed corners and intersections (shown in plan view).

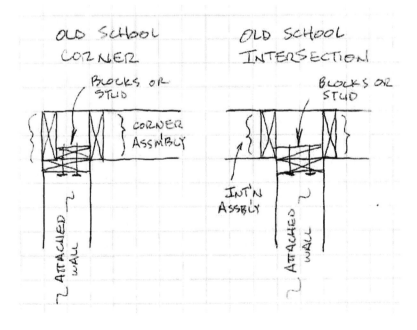

What's wrong with these?

To start, there's too much wood consumed. The assemblies use three studs each. Okay, sometimes the middle "stud" would be several blocks, which is a little greener. But could we do better? From a structural standpoint, how many studs are really required?

Let's look at gravity loads first. Does a corner or intersection carry more gravity load than other places along the wall? The answer is no, unless there's a point load coming down there from a beam or post above. Let's say there *is* a point load from above. How much vertical load can two studs carry? How much can three carry?

Studs in a corner or intersection are braced in both directions against buckling so they behave as "short" compression members. Using "Column, Post, Stud Calculator"™ by ConstructionCalc™, setting unbraced length to 1', I determined the following:

Studs Braced In Both Directions, Vertical Load Carrying Capacity:

- 2, SPF Stud Grade studs = 12,600 lbs.
- 3, SPF Stud Grade studs = 18,900 lbs.

That's a lot of load. Most beams in stick-framed construction bring less than half of that. Not to mention, most beams in stick-framed construction do not bear at corners or intersections.

The point is that if we can figure out how to use even two studs in a corner or intersection, we're going to be okay structurally.

What about wind load? Because the walls are sheathed inside and out, and because two walls join at 90-degrees, corners and intersections are fully braced against wind load. Wind blowing on these elements is absorbed in the plane of the perpendicular wall, eliminating the possibility of bending stress in the assembly itself. Thus we are able to ignore wind load in our analysis.

What about seismic load? In stick-framed construction wind forces on a stud, corner, or intersection will always be greater than seismic forces. So wind controls the design and we need not consider seismic forces in these analyses. Certainly this is not the case for shear walls and their elements but that's another topic.

The second problem with our old school corners and intersections is that where we're using too much wood, we're using too little insulation.

The third problem would only apply to corner assemblies and only when shear forces exceed 350 plf. In that case code requires double 2x or 3x boundary members but there's only a single 2x available.

Following are a couple of green suggestions (plan view). The corner assembly uses two studs and leaves maximum space for insulation. The intersection uses only one stud and creates no wood thermal bridge. Very green.

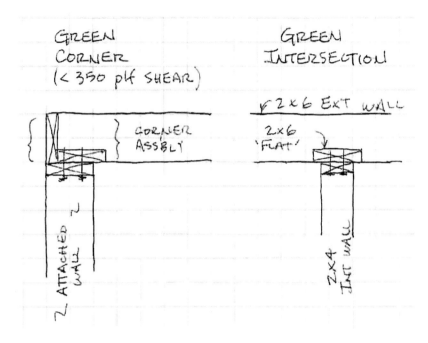

GREEN
CORNER
(< 350 plf SHEAR)

CORNER
ASSBLY

2 ATTACHED WALL

GREEN
INTERSECTION

← 2×6 EXT WALL

2×6
'FLAT'

2×4
INT WALL

What about the corner case where shear exceeds 350 plf? The above green corner will not work because there needs to be a double 2x or single 3x boundary member in the corner but the corner stud is only a single 2x. One solution would be to replace the corner stud with a 3x. Here is another option that isn't quite as green.

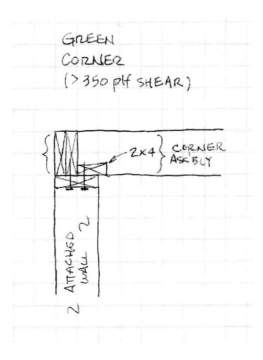

GREEN
CORNER
(>350 plf SHEAR)

2x4 } CORNER ASSBLY

ATTACHED 2 WALL 2

The shear requirement boosts the number of studs back to three, though one of them could be a 2x4 of low quality because its only purpose is drywall backing.

Shear Walls, Holdowns, and Anchor Bolts

(Author's note. A deeper treatment of the concepts in this section can be found in my book, "Basic Structural Concepts for the Non-Engineer".)

Perhaps the worst green framing offender is the shear wall. While it's true that engineers pretty much have the market cornered on their design, every builder and architect should understand what's involved so that tough questions can be asked when too many tons of materials show up on plans. Recall from Chapter 1 that engineers have little incentive to produce efficient designs. To them, more is safer, and safer equates to less liability.

Here are the basics.

Wind and earthquake affect structures similarly – they apply (for the most part) sideways forces. Engineers calculate the worst case, wind or seismic, and design shear walls, moment frames, etc. to resist that loading. For the remainder of this book, when I say "lateral load" I mean the worst case wind or earthquake load, whichever controls.

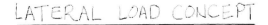

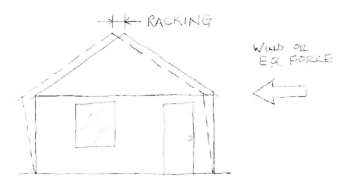

Applied lateral loads are collected in horizontal diaphragms - roofs and floors. Those diaphragms then transfer the lateral loads to vertical shear-resisting elements below. Whenever I say "vertical element" I mean either a shear wall or moment frame (see Definitions at the front of this book for more on these terms.)

Thus racking and sideways collapse are resisted by vertical elements, usually shear walls in stick-framed construction.

Shear Transfer and Load Path. In order for a vertical element to function, loads must actually get to it. As I said in the above paragraph, lateral loads are collected in horizontal diaphragms and then transferred to vertical elements. That *transfer* can be difficult to achieve.

Assuming lateral load does get to the top of a vertical element, it has to get back out – another *load transfer*. Ultimately all loads are resisted by the earth, thus the final load transfer occurs at the foundation.

The transfer of lateral loads into and out of vertical elements starting at the roof and ending at dirt is called "load path." Like the weak link in a chain, any breach in the load path compromises the entire system.

Engineers frequently use framing clips, and lots of them, to transfer lateral loads from one element to another. Here is a green opportunity.

In the following sketch we see how lateral load is typically transferred from a roof diaphragm to a gable end wall. There's not a lot we can do to green this detail except perhaps get rid of the framing clips at the bottom chord of the gable truss. If said clips are spaced at, say, 24", there could easily be 20, total, and wouldn't it be nice if we didn't have to install them.

Their purpose is to transfer in-plane lateral loads out of the gable truss and into the shear wall below. If a horizontal joint in the wall sheathing (OSB or plywood) was located at exactly that location, the clips would be necessary. However, if we locate our sheathing horizontal joint elsewhere, there will be

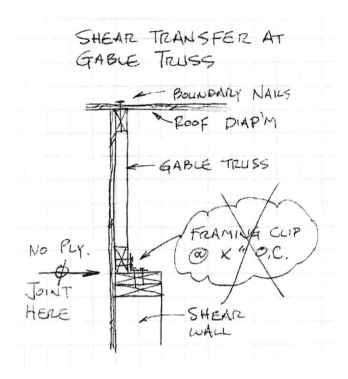

SHEAR TRANSFER AT
GABLE TRUSS

BOUNDARY NAILS

ROOF DIAP'M

GABLE TRUSS

FRAMING CLIP
@ X " O.C.

NO PLY.

JOINT
HERE

SHEAR
WALL

continuous sheathing over the truss-wall intersection, which will transfer the shear between those two members just fine.

Where should we locate the horizontal joint? It really doesn't matter as long as solid framing exists behind so that edge nailing can be provided. A good place would be 1-1/2" above or below the bottom chord–top plate intersection. In either location the sheathing joint would have at least 3" (measured vertically) of backing for edge nailing both upper and lower pieces of sheathing.

Now let's look at a typical eaves detail.

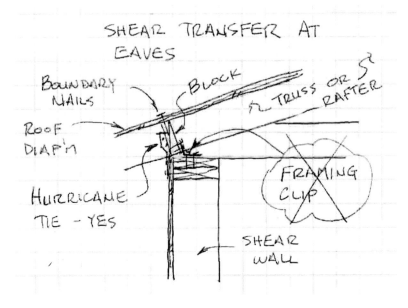

SHEAR TRANSFER AT EAVES

BOUNDARY NAILS

BLOCK

TRUSS OR RAFTER

ROOF DIAP'M

HURRICANE TIE - YES

FRAMING CLIP

SHEAR WALL

Again, it would be nice to get rid of the framing clips attaching the blocking to the top plate. In a typical house, there might be 50 of these.

Lateral load is transferred from the roof to the blocking via the roof boundary nails. From there the load must get into the shear wall. There is little opportunity to run the wall sheathing up, covering the blocking similar to our strategy with the gable truss. So we need a mechanical attachment of some sort. But do we need both a hurricane tie and a framing clip?

Generally, no. A hurricane tie is usually provided connecting the truss or rafter heel to the wall, its main purpose to keep the truss / rafter held down in the event of uplift from wind. However, hurricane ties also provide resistance to sideways forces as well. Here are a couple common ties and their strength values.

- Simpson H2.5. Uplift capacity = 415 lb. In-plane lateral capacity = 150 lb.

- Simpson H1. Uplift capacity = 585 lb. In-plane lateral capacity = 485 lb.

Thus H1s are good for nearly any residential application and H2.5s work when lateral loads are light to medium.

So the load path is: roof diaphragm to blocking, which butts against truss / rafter heels, which are connected to the shear wall with hurricane ties. Mission accomplished.

Holdowns. If I have a pet peeve it is too many holdowns. They're expensive and time-consuming to install. Sometimes they're necessary but many, many times they're not.

The following sketch shows the concept of lateral loads on a multi-level structure with a bunch of narrow shear wall panels. Notice that the lower level takes load tributary to itself plus load from the floor(s) above. This is why there are usually more holdowns on the lower floors of any multi-story building.

Why are holdowns so overused? Here are several reasons:

- It's a lot easier and faster for an engineer to simply throw in a bunch of holdowns than to sharpen his pencil and do shear wall analysis correctly.

- Too many holdowns never poses a safety risk. Too few might. Engineers love things that minimize their liability. Heck, they're not the ones paying for all those holdowns anyway.

o To minimize holdowns requires a certain degree of creativity during analysis. Engineers are not known for their abundant creativity.

UPLIFT AND HOLDOWNS

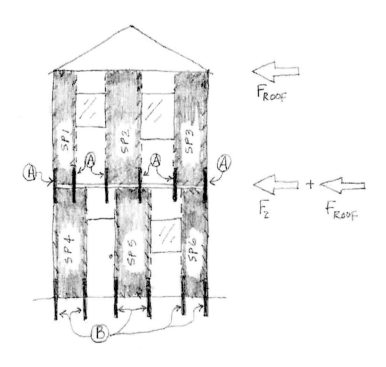

Ⓐ FLOOR-TO-FLOOR HOLDOWN STRAP
Ⓑ HOLDOWN ANCHOR TO FOOTING

In lateral analysis, doors and windows are assumed as floor-to-floor "holes" in the building. Resistance to wind and earthquake loading comes only from full-height, code-compliant shear wall "panels". When lateral load is applied to a panel, if there isn't enough dead weight to hold it down, it would tend to tip as shown in the following sketch.

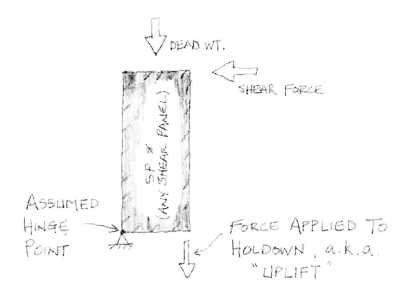

Actually, shear forces can go either left or right depending on which way the wind blows or the ground shakes, which accounts for holdowns on each side of most shear panels.

The rub comes in determining the resistive dead weight. The more dead weight, the less likely the panel will tip. If the panel has so much weight holding it down that it can't tip, it doesn't need a holdown. Lazy engineers might only count the weight of the panel itself, which isn't much. Kind-of-lazy engineers might also include the weight of the roof or floor bearing on the panel. Green engineers, however, will count ALL the dead load holding the panel down.

Here's a sketch showing all the dead load that can be counted holding down a shear panel.

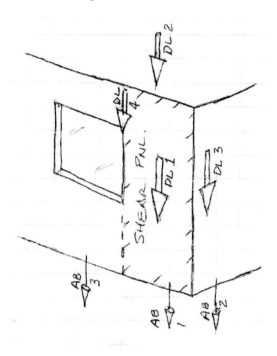

DL1: The weight of the panel itself.

DL2: The weight of roof or floor or both directly bearing on the panel.

DL3: The weight of the adjacent wall. This only helps hold down one side and can only be counted if the two walls are well-connected.

DL4: The weight of roof or floor or both above the adjacent window which is brought down to the edge of our shear panel via a header and trimmer.

AB1: An anchor bolt within ~12" of the edge of our shear panel.

AB2: An anchor bolt on the adjacent wall within ~12" of our shear panel.

AB3: An anchor bolt within 6' of AB1 (or closer, depending on specified spacing).

Other elements, though not shown in the above sketch, can help hold a shear panel down too, including, beam(s) that happen to bear on a shear panel, or an intersecting interior wall which is connected to a shear panel.

Some engineers might question the above because it's hard to quantify exactly how much certain dead loads and / or anchor bolts contribute. I agree, it is hard to make those calculations, however, the alternative is to not count them at all, which results in artificially high overturning forces and lots of expensive, non-green holdowns. I have come up with conservative estimates that I use in my engineering practice. Any competent engineer should do the same.

The other main variable in calculating the need for holdowns is the *applied shear force*. Calculating those is well beyond the scope of this book, but understanding the concepts so as to minimize hardware is understandable by everyone.

When wind or earthquakes occur, how in the world do engineers know where those forces are distributed throughout the building? Do we have a conversation with the Plans?

"Okay, shear walls A, E, and R, you guys take 3,200 lbs. Panels 11, 9, 22, and 3, you only need to worry about 2,734 lbs. Ready, *break!*"

Not exactly. The building code tells us how to apportion lateral loads. But does that mean the building will truly behave as the code dictates? No way. In fact, I bet if ten different engineers were given identical sets of house plans, ten different sets of lateral forces and ten different lateral resisting system designs would be produced. Then when the earthquake struck, not one of the ten would behave as intended.

There's just too much going on in a building to make any sort of precise prediction, such as: the role of non-shear-walls (walls not counted in lateral design); the roll of windows and doors (they're not counted but they do help); the role of drywall on the other side of a plywood wall (not counted in the design); the roll of mullions and king studs (behave as moment frames but are not counted); the roll of cladding on exterior walls (not counted but it contributes); and so on.

The point is engineers can be creative when performing lateral design. They can, to a certain extent, tell the plans where the loads will go. Of course, it all needs to make static and dynamic sense, but there is some flexibility. Here's how I go about it.

Tim's Lateral Design Criteria

1. I try to keep loads in any particular section of wall low. This minimizes holdowns, 3x boundary framing and other expensive shear wall parts.

2. I recognize that a good lateral resisting system is somewhat symmetrical, i.e. there should be similar amounts of shear walls or moment frames on all four sides of the building.

3. I try to use *exterior* shear walls (as opposed to interior) as much as possible, keeping point 1., above, in mind at all times.

4. When I need to use interior shear walls to accomplish 1. above, I do, but with the following provisions.

5. When using interior shear walls, I do my best to keep them constructed of drywall, not expensive plywood or OSB. Drywall shear walls are code-compliant, they just can't take as much force as other materials.

6. A big problem with interior shear walls is getting the load into and out of them, i.e. satisfying load path. This is particularly difficult when the wall is perpendicular to roof trusses above. (This is why I have big heartburn with the IRC's prescriptive lateral design requirements. Interior shear walls are mandatory by that code, but there is very little explaining how correct load path is to be achieved.)

7. Another big problem with interior shear walls occurs when they require holdowns. What are those holdowns secured to below? If not a footing, there will be complications.

To summarize, holdowns can be kept to a minimum by, 1) Counting ALL the dead load and other mechanical means already there holding them down, and 2) Selecting a smart lateral resisting system which spreads lateral load around, keeping forces low in any particular element.

Shear Walls Themselves. Different types of shear walls have different lateral strengths. In general, the greater the strength of a wall, the more expensive it is to construct. We're trying to build green, so we want to design the least expensive, least material-intensive shear walls we can.

Here are several shear walls I typically spec and their strengths:

- o **PW1**: 7/16" plywood or OSB sheathing, one side, 2x boundary members, 8d nails at 6" edges, 12" field, good for **260** plf.

- o **PW2**: 7/16" plywood or OSB sheathing, one side, 3x boundary members, 8d nails at 4" edges, 12" field, good for **380** plf.

- o **PW3**: 7/16" plywood or OSB sheathing, one side, 3x boundary members, 8d nails at 3" edges, 12" field, good for **490** plf.

- o **PW4**: 7/16" plywood or OSB sheathing, both sides, 3x boundary members, 8d nails at 4" edges, 12" field, good for **760** plf.

- o **PW5**: 7/16" plywood or OSB sheathing, both sides, 3x boundary members, 8d nails at 3" edges, 12" field, good for **980** plf.

The above assumes studs at 16" spacing and all edges of sheathing are blocked. I didn't list them here but there are different, slightly lower values for similar construction with studs at 24" spacing.

With plywood and OSB shear walls, all edges must be blocked. However, blocking is not required for most roof and floor diaphragms, nor for certain drywall shear walls. Blocking costs money so I avoid it when I can.

Here are examples of interior, drywall shear walls I often use:

- o **GW1**: 1/2" gyp board, both sides, unblocked, 2x boundary members, studs at 24" OC, 5d nails or screws at 4" throughout, good for **110 plf**.

- o **GW2**: 5/8" gyp board, both sides, unblocked, 2x boundary members, studs at 24" OC, 6d nails or screws at 4" throughout, good for **145** plf.

- o **GW3**: 1/2" gyp board, both sides, blocked, 2x boundary members, studs at 16" OC, 5d nails or screws at 4" throughout, good for **150** plf.

- o **GW4**: 5/8" gyp board, both sides, blocked, 2x boundary members, studs at 16" OC, 6d nails or screws at 4" throughout, good for **175** plf.

There are many other combinations with different strengths – too many to list here – but you get the idea. Engineers have a lot from which to choose.

When I design, I try to keep the force in exterior shear walls below 350 plf. That is the threshold at which 3x boundary members become required. If I have to use interior shear walls I try to use options that allow 24" stud spacing and no blocking.

Sometimes a structure does not lend itself well to this approach, leaving little room for creativity or green design. But most times, especially with stick-framed construction, that's not the case – creativity is possible.

Anchor Bolts. Anchor bolts transfer lateral loads from the bottom of shear walls to concrete foundations. Without anchor bolts, structures are easily knocked off their foundations in earthquakes or wind storms.

The IBC requires, at a minimum, 1/2" diameter anchor bolts spaced no greater than 6' OC. This is one of a very few areas of code that I tend to exceed.

I've seen a lot of earthquake damage over the years and have noted that one of the most common areas of failure is at anchor bolts. It's not that the bolts shear off, its more that the wood ruptures around them or the host concrete cracks around the embedded portion. When I design anchor bolts I tend to err on the conservative side. But still there are a couple "tricks" I employ.

Even though code allows it, I have not specified a 1/2" diameter anchor bolt in over 20 years. I always spec 5/8". Why? Because that measly 1/8" difference yields a significant increase in holding power.

- o 1/2" diameter anchor bolt, in-plane strength ~ 745 lbs.
- o 5/8" diameter anchor bolt, in-plane strength ~ 980 lbs.

That's over 30%. I find it more efficient to use fewer 5/8" diameter bolts.

When engineers calculate anchor bolt spacing, frequently they only count shear panels and not the portions of walls under windows. I ask why not? In order for any shear panel to slide, the entire wall has to go along too. Granted, you can't count on a widow to resist racking, but you can count the bottom plate connection below the window to keep the wall from sliding off the foundation.

Headers, Beams, King Studs, and Trimmers

In Chapter 2 we looked at quite a few examples of oversized headers and too many king studs and trimmers. Now let's get to the how and why.

The variables we used to design roof beams are the same ones we use to design headers. A header is a beam, after all, just one that spans a door or window opening.

Here are the main criteria that go into the design of any beam:

o Span

o Loading

o Depth for clearance height or architectural purposes

o Deflection criteria

Let's design a header over a sliding glass door.

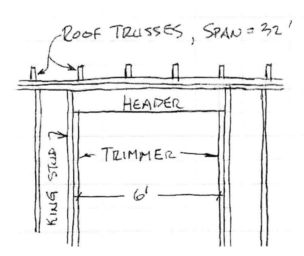

Span. We're given a rough opening of 6' which is the span of our header.

<u>Loading</u>. This header supports roof trusses. We'll use the same loads as we did for the roof chapter examples:

Snow: 25 psf

Live: 16 psf, but snow controls.

Dead: 15 psf. This does not include the self-weight of the header. We'll add that in ProBeam™.

There isn't much opportunity to green the design by tweaking loads. They are prescribed by code.

<u>Tributary Width</u>. Load is applied to our header by roof trusses. The tributary width of the trusses is half their span plus any overhang. Let's say we have 1' overhang, thus the tributary width is, 32'/2 + 1' = 17'

<u>Depth</u>. The header's depth (the vertical dimension) can be anything that allows a sliding glass door to fit beneath it. Here is a green opportunity. By using exactly the right depth we can avoid any extra fill-in framing. Let's say that depth is 11-1/4".

<u>Deflection Criteria</u>. From Table 1604.3 in the 2006 IBC, deflection is limited to l/360 for live load only, and l/240 for total load (live + dead). Live in this case is snow load.

<u>Calc it</u>. Here is the general input.

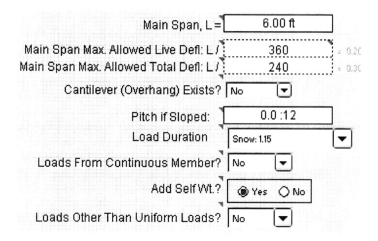

And here is the loading input.

r Full Length of Member		Tributary
	Live, psf	Dead, psf	Width, ft
ot including snow)		15 psf	17.00 ft
Roof Snow (only)	25 psf		17.00 ft
Floor 3 Loads			

Done. Now let's check our options.

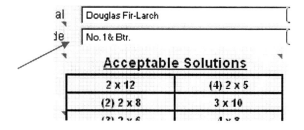

First, sawn wood. We'd love to use a 2x12 – it's exactly the right height and leaves lots of space for insulation. But we find that with our default species and grade, Doug Fir No. 2, a 2x14 is required. If we change the grade to No 1 and Better, a 2x12 works.

That's great if we can find that particular piece of wood at the lumber yard. Here's an option I like just as well.

1.55E Timberstrand LSL	
1-3/4" x 9-1/4"	(3) 1-3/4" x 9-1/4"
(2) 1-3/4" x 9-1/4"	Slf Wt= 6

This material comes 11-1/4" as well as 9-1/4". By inspection we know that the larger size, though not specifically shown in the table, works.

Certainly we could have used any of the options shown, such as 2, 2x8s, however, with the installation of framing spacers to correct for height, and less available space for insulation, our green goal would have been compromised.

Speaking of framing spacers, if we use a 2x header exactly the right height, shifted to the inside of the wall, to what do we nail exterior sheathing at the bottom of our header? Here are a couple options.

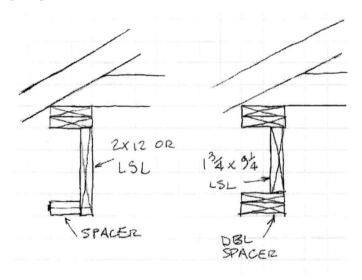

The detail on the left is greenest but requires ripping a spacer (or spacer blocks – it doesn't have to be 6' long). The other detail doesn't require a ripped spacer but it's not as green, and also, the height is not exactly 11-1/4".

So to summarize, there are many ways to skin this cat – some we didn't even mention. The method you choose in the end may come down to personal preference – i.e. how you like to frame. Regardless, if you know how to calc the header, you control that choice.

Trimmers. How many trimmers are needed to support our header? One on each side? Two? None? The answer, in part, has to do with potential crushing of the header where it bears on the trimmer. If there isn't enough trimmer width for the header to rest on, the bearing surface of the header could crush. Let's go back to ProBeam™ and select as Final Member the 2x12 Doug Fir No 1 and Better. Just below the Final Size we're given the minimum required bearing length of 2.19".

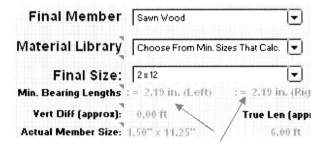

A 2x trimmer is 1.5 inches wide – not enough. If we're dead set on using a Doug Fir header, we'd either have to use double trimmers on each side (bearing length = 3", not green at all) or for our header use 2, 2x8s or 3x10, or 4x8 (would reduce the required bearing length to 1.5" - try it).

But let's see how much bearing area is required with the LSL option.

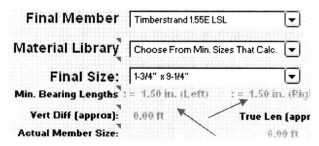

Now we're talking. The LSL is considerably stronger and more resistant to crushing than is Doug Fir. It only requires 1.5 inches of bearing length, so now a single 2x trimmer works. *Yes*.

The other part of trimmer design is calc'ing its compressive capacity. A trimmer, after all, is nothing more than a post which happens to support a header. Calc'ing one is easy with the right software.

The reaction at the end of the header is the applied gravity load on the trimmer. Since the load over the header was uniform we would expect the reactions at each end to be the same, which they are. From the results section of ProBeam™:

| | **Reactions** | |
Maximums	R₁ - Left	R₂ - Right
Live Load:	1,275 lb	1,275 lb
Dead Load:	783 lb	783 lb
Total Load:	2,058 lb	2,058 lb

The above live and dead loads are what we input into Column, Post, Stud Calculator™. First, general input.

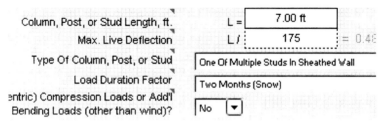

… and here is our loading.

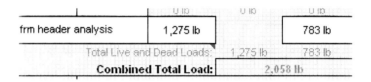

What about wind load? Do trimmers resist wind load? No, because they do not extend from top plate to bottom plate. So we leave that section blank.

Done. Here are some choices.

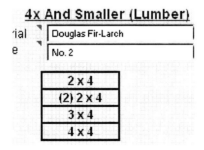

We could use a 2x4 or anything larger.

Calc'ing this trimmer was largely academic – we pretty well knew that a single, 2x works when supporting a 6' long header. It was good practice, however, for the case when the header or beam gets large and we're not sure how many trimmers are really needed. We'll get to one of those examples shortly.

King Studs. How many king studs are required on each side of our header? Standard practice would call for one, and I agree that is the right number.

What does a king stud do exactly? If the trimmer takes the entire gravity load from the header, then the king stud's main function is to take lateral load applied perpendicular to the wall (out-of-plane.) It might also pick up a small amount of gravity load from the next truss over from the header. Let's calc it. Here's the general input.

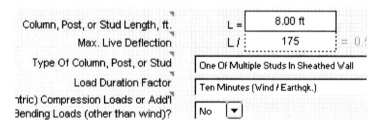

Column, Post, or Stud Length, ft.	L =	8.00 ft	
Max. Live Deflection	L /	175	= 0.!
Type Of Column, Post, or Stud	One Of Multiple Studs In Sheathed Wall		
Load Duration Factor	Ten Minutes (Wind / Earthqk.)		
ntric) Compression Loads or Add'l Bending Loads (other than wind)?	No ▼		

We changed the length to 8' and changed the load duration factor to reflect the worst case, wind. Here is the gravity load.

	Live, psf	Dead, psf	x Length, ft	x Width, ft.
w)		15 psf	1.00 ft	17.00 ft
y)	13 psf		1.00 ft	17.00 ft
ds				

We remember from calc'ing a stud previously that the worst case loading came from full wind + 1/2 snow. Regarding tributary area, if the next closest stud was 24" away, the king stud would support an area of roof half that distance, 1', times half the truss' span + 1' overhang, 32'/2 +1' = 17'.

Here is the wind loading.

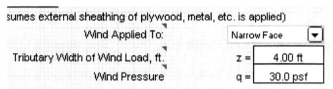

The Tributary Width of wind load is half the distance to the other king stud + half the distance to the next stud on the non-door side, 6'/2 + 2'/2 = 4'.

We're using the same wind pressure as previously, a "medium" wind load of 30 psf.

Done. Here are some options.

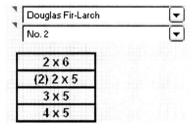

A 2x6 Doug Fir No. 2 makes it, but by how much?

This member makes it by: 16.9%

Controlling Factor: Combined Bending / Compressive Stresses

By 16.9%. That's efficient but there may be room for a lesser grade, such as Hem Fir or Spruce Pine Fir. With this software, checking those options only takes a couple clicks. Try it.

Something else to keep in mind, this 2x6 king stud doesn't have a lot of overdesign for a measly 6' wide sliding glass door. Do you think a single 2x6 king stud would calc for a 12' wide garage door? I bet not, not in an area of significant wind. What about if the ceilings in the house with our sliding glass door were 10'? To check that requires one input change – the Length. I tried it and found that a 2x6 no longer works, not even a No 1 and Better. Double king studs are required. What does this tell you about the greenness of 10' ceilings?

Header At Gable Wall. Let's say our 6' sliding glass door was located in a wall with a gable truss over it. Intuitively we know that a much smaller header will be called for because the loading is much, much less.

Let's try something radical, something green to the extreme - *no header at all*. This is what it might look like.

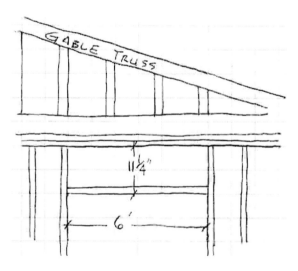

Notice there are no trimmers either. Can we even do this? Will code allow it? As long as gravity and lateral loads are taken care of and proper backing for drywall and exterior sheathing exists, yes we can.

Here's how. First, what will take the gravity load from the gable truss if we don't have a conventional header? Can the gable truss itself span 6'? I bet it can easily. However, the truss manufacturer probably wouldn't allow it so we need to provide something under it. What about our double top plate? Let's assume we don't allow a splice in either of the double plates within the 6' span of our door opening. And let's assume the two plates are nailed together so they act as a single piece of wood 5.5" wide x 3" tall (16d at 6" OC, staggered would do that nicely.) Let's calc it.

The general input is the same as when the header was in the side wall. The uniform load section has one input change, however. Tributary width is half the distance from the gable truss to the next truss in, plus overhang: $2'/2 + 1' = 2'$.

Before investigating a double top plate, let's take a look at standard results. In the "Lumber 4x and Smaller" section, we find that a normally-oriented 2x4 Doug Fir No. 2 works (all results in Part 3 of ProBeam™ assume strong axis orientation), so we could use that and be done. But how green is it? Not so much. To make that work we'd have to install cripples and also use an "L" header to provide backing. It might look something like this.

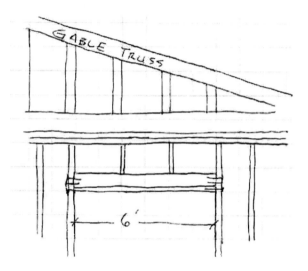

Notice we're still not using trimmers. How is that possible? The reactions at the ends of our header are only 252 lbs. each. (from the Results section of ProBeam™). A 16d nail can resist about 150 lbs. So two nails per end, king stud to header, works. I'd probably use 3 or 4 for good measure.

This is all well and good but we can do better. Let's calc the double top plate as a header: 2, 2x6s oriented flat.

To do this we use the Custom Member feature in ProBeam™. That section is located just above the results section, and is normally hidden. Unhide it via the Miscellaneous dropdown in the blue header at top, or select "Custom Member" from the Final Member dropdown in Part 4.

First we select the Shape from the dropdown. Then we input the Width and Height. Here's what that looks like.

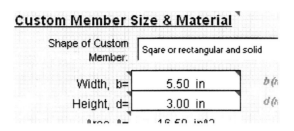

Custom Member Size & Material

Shape of Custom Member:	Sqare or rectangular and solid
Width, b=	5.50 in
Height, d=	3.00 in

Now we need to input allowable strength values. Most people, myself included, don't walk around with allowable stresses of wood in the forefront of their minds, so fortunately there are quite a few in the various libraries of ProBeam™. The Typical Values dropdown has exactly what we're looking for.

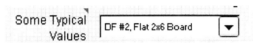

Some Typical Values | DF #2, Flat 2x6 Board ▼

Selecting that automatically pastes the correct allowable stresses in the cells immediately above.

Done. Results are shown in the section to the right…

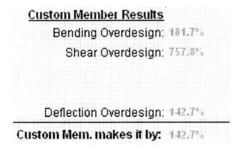

Custom Member Results
Bending Overdesign: 181.7%
Shear Overdesign: 757.8%

Deflection Overdesign: 142.7%

Custom Mem. makes it by: 142.7%

Excellent – our double 2x6 top plate is plenty strong.

Now what about lateral loads? Regarding this wall being a shear wall, this sliding glass door, like all doors and windows, is not counted in resisting in-plane lateral loads. So we're fine

---

there. Regarding out-of-plane wind loads, we already know that the king studs do all the work. And we already calc'd 8' tall king studs with more gravity load than these in the previous example. So these king studs are okay by inspection.

Bottom line, our exceptionally green "non-header" is okay.

Will this non-header work for any gable end window or door? Absolutely, so long as the span and loading do not exceed this example's, and the double plates are not spliced over the door / window opening. How much wood did we just save? Extending this concept over the entire house, maybe a small tree's worth.

Heavily-Loaded Beam. Now let's examine a beam with some real load on it, a garage door header supporting roof and floor loads.

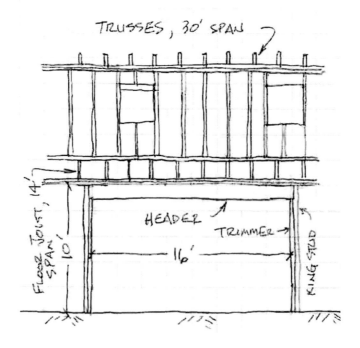

Most framers intuitively know that this will be a big beam and the trimmers will be more than just one on each side. But how big the beam and how many trimmers? And also, will a single king stud work?

In cases like this I've seen a 6x12 Doug Fir header with 6 or 8 trimmers used per side, with only one king stud. Let's do it right this time, let's calc it.

Step one is to calc the beam. Here's the general input.

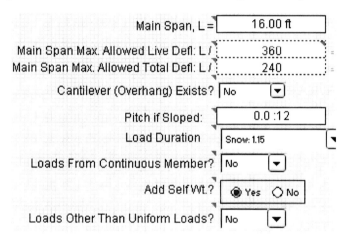

Main Span, L =	16.00 ft
Main Span Max. Allowed Live Defl: L /	360
Main Span Max. Allowed Total Defl: L /	240
Cantilever (Overhang) Exists?	No
Pitch if Sloped:	0.0 :12
Load Duration	Snow: 1.15
Loads From Continuous Member?	No
Add Self Wt.?	Yes No
Loads Other Than Uniform Loads?	No

The only surprise might be the Load Duration. I used Snow: 1.15 because the worst case live load will occur when snow is on the roof, and this beam supports a big chuck of roof load. Generally, you size structural members for the worst case live load and select the Load Duration factor corresponding to that.

Next the loads. Our header supports three sets of loads. Starting at the top is the roof. We'll use the same snow and dead loads as before. Tributary width is 30'/2 + 1' overhang = 16'. Next is the exterior wall. This is purely a dead load of 10 psf. The tributary width for a wall on a beam is the wall's height, in this case 8'. Last is the 2nd floor. Residential floor live load is 40 psf. For dead load let's assume carpet flooring on top and drywall ceiling on bottom: 15 psf. Tributary width is half the span: 14'/2 = 7'. Here's what it looks like in ProBeam™.

Full Length of Member			Tributary
	Live, psf	Dead, psf	Width, ft
ɔt including snow)		15 psf	16.00 ft
Roof Snow (only)	25 psf		16.00 ft
Floor 3 Loads			
Floor 2 Loads			
Floor Loads	40 psf	15 psf	7.00 ft
Wall Dead Load		10 psf	8.00 ft
ɔɑd ɑnd trib. width			

Done. Now let's check some alternatives.

The smallest sawn Doug Fir No.2 beam is a 6x24. Whoa, that's big. The guy who used a 6x12 for this header was way off. Dangerously so.

Here are our glu-lam options.

Glued Laminated Members	
24F-V4 (DF/DF) ▼	
Acceptable Solutions	
2.5" x 19.5"	5.125" x 15"
3" x 18"	6.75" x 13.5"
3.125" x 18"	8.75" x 12"
5" x 15"	

The 5x15 or 6.75x13.5 look promising. Our final determination may have something to do with the size of studs in this wall because we don't want our beam wider than the wall itself.

Also of interest are PSL alternatives.

2.0E Parallam PSL

	5-1/4" x 14"
2-11/16" x 18"	7" x 14"
3-1/2" x 16"	Slf Wt= 35

The 5-1/4x14 looks like the most efficient one yet.

Let's say this client has an SUV with a ski rack on top and he needs as much vertical clearance as possible. We check the garage door opener height and determine that we can live with a 14" tall header but no more. If we use the PSL we're good with height, and the studs can be 2x6. Done deal. Let's select that as our Final Member and check bearing length and reactions.

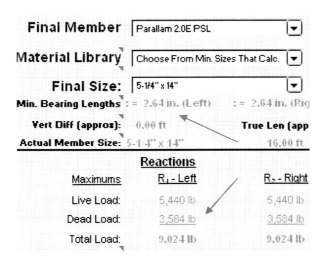

Final Member	Parallam 2.0E PSL ▼
Material Library	Choose From Min. Sizes That Calc. ▼
Final Size:	5-1/4" x 14" ▼

Min. Bearing Lengths : = 2.64 in. (Left)   : = 2.64 in. (Rig

Vert Diff (approx): 0.00 ft   True Len (app

Actual Member Size: 5-1/4" x 14"   16.00 ft

Reactions

Maximums	R₁ - Left	R₂ - Right
Live Load:	5,440 lb	5,440 lb
Dead Load:	3,584 lb	3,584 lb
Total Load:	9,024 lb	9,024 lb

Before we move on it's noteworthy that the dead load at each end of our header is over 3,500 lbs. Do you think holdowns are needed there? Probably not, provided the engineer doing lateral analysis counts this significant dead load.

Trimmers. Minimum required bearing length is 2.64 inches so we know at least 2, 2x trimmers are needed per end (bearing length of two trimmers = 1.5"+1.5"=3"). But do we really need six or eight to take the load? Let's calc it.

Here's the general input from Column, Post, Stud Calculator™.

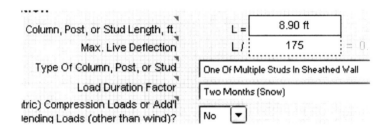

Keep in mind trimmers do not resist wind load, king studs do that. We use the same Load Duration Factor as when sizing the header - Snow.

Our gravity loads come from the header reactions.

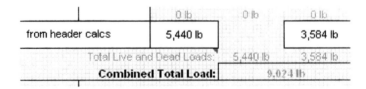

The wind load section is left blank.

Done. Let's check results.

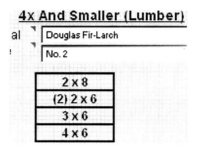

4x And Smaller (Lumber)

| al | Douglas Fir-Larch |
| ! | No. 2 |

| 2 x 8 |
| (2) 2 x 6 |
| 3 x 6 |
| 4 x 6 |

Two, 2x6 trimmers work. And not only that, checking their efficiency in the results section we see that they make it by a very comfortable margin, 33.2%. So the guy who used six trimmers per side wasted 4 x 2 sides = 8 trimmers.

King Studs. As we've said previously, trimmers take the gravity load, king studs take the out-of-plane wind load.

When the garage door is closed, wind blowing on it is transferred to the king studs on each side. So the tributary width of wind load on each king stud is half the width of the door + half the distance to the next stud. Assuming studs are spaced 16" OC, tributary width is: 16'/2 + 1.33'/2 = 8.6'.

Here's the input from Column, Post, Stud Calculator™.

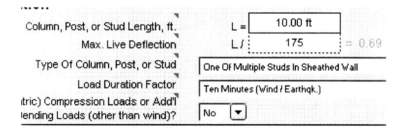

Column, Post, or Stud Length, ft.	L =	10.00 ft	
Max. Live Deflection	L /	175	= 0.69
Type Of Column, Post, or Stud	One Of Multiple Studs In Sheathed Wall		
Load Duration Factor	Ten Minutes (Wind / Earthqk.)		
tric) Compression Loads or Add'l lending Loads (other than wind)?	No ▼		

Note the Length and Load Duration are different than the trimmer example.

What about gravity loads? We could include a small amount of gravity load, but with studs spaced at 16" OC, it is safe to assume that all gravity load will be taken by trimmers and the next adjacent stud. So we leave the gravity load section blank.

We'll use the same "medium" wind pressure as before, and the wind tributary width that we calc'd above.

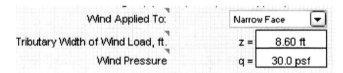

Done. We see that nothing shows up in 2x6 but 2, 2x8s work.

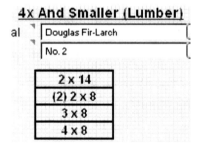

This software program doesn't support three plies so we're left wondering whether 3, 2x6's would work. Here is a simple equation that gives a very close approximation.

$(H^2 * \text{NO. PLIES}) / (h^2 * \text{no. plies}) - \text{overdesign}/100 < 1$

Where, H = height of member that calcs.

NO. PLIES = number of plies of the member that calcs.

h = height of member you want to use.

no. plies = number of plies of member you want to use.

overdesign = the percentage by which the member that calcs makes it.

<1 means the answer must be less than or equal to 1.0 or you need another ply.

In our case, $(7.25^2*2) / (5.5^2*3) - 24.5/100 = .92$ which is less than 1.0, so okay.

Thus using 2x6, Doug Fir No. 2, we need 3 king studs and 2 trimmers on each side of our header. Here is a sketch of what that would look like.

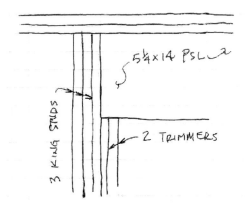

Interior Headers, Trimmers, and King Studs. So far we've looked at headers, trimmers, and king studs on exterior walls. But what about *interior* walls? Do interior headers ever bear weight? Need we worry about lateral loads?

The simplest case is when a door or window exists in a non-load-bearing wall. An example of such a wall would be one parallel to roof trusses or floor joists above. In this case no header is required. The following sketch shows what this might look like. Notice that there is no header, no cripples, and no trimmers. The double top plate could even have been a single. No calcs required. Very green indeed.

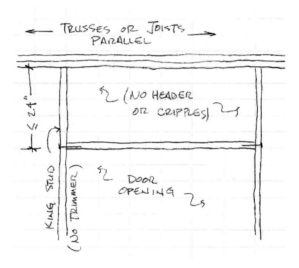

If the interior wall *is* load bearing, headers and trimmers must be designed as we did previously for exterior walls. Here is an example.

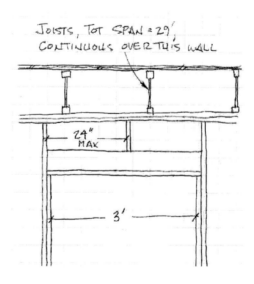

Let's calc this header. Here is the general input from ProBeam™.

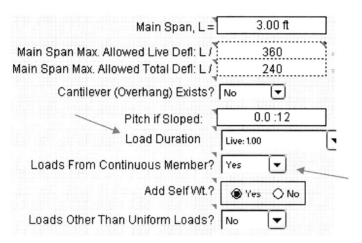

Notice a couple differences from previous analyses. First, Load Duration is "Live 1.00" rather than "Snow" because this header doesn't support a roof - it supports a floor only. Second, I have selected "Yes" for "Loads From Continuous Member?" because the floor joists are not spliced over this wall. This selection increases the applied uniform loads 15%. But what if the joist supplier provides 15' joists and a splice *does* wind up over this wall? If the joists are spliced, that 15% extra never really exists, thus the header will be oversized, which, while not particularly green, is conservative. If in doubt, select "Yes" – it's always conservative.

Live load for residential floors is 40 psf. For dead load, we'll assume carpet on the top and gyp ceiling on the bottom, 15 psf. Tributary width is half the span of the floor joists on one side of the wall plus half the span on the other side. This always turns out to be half the distance between the two adjacent bearing walls, in our case, 29'/2 = 14.5'.

'ull Length of Member	Live, psf	Dead, psf	Tributary Width, ft
including snow)			
:oof Snow (only)			0.00 ft
Floor 3 Loads			
Floor 2 Loads			
Floor Loads	40 psf	15 psf	14.50 ft

Done. Let's check results.

al	Douglas Fir-Larch
e	No. 2

Acceptable Solutions

2 x 8	(4) 2 x 4
(2) 2 x 5	3 x 6
(3) 2 x 5	4 x 5

We could use a 2x8. Or 2, 2x6s would work. We could even throw in a 4x6 if we had one laying around that we wanted to use up. I like the 2x8 used in a "T" fashion, like is shown on the next page.

A warning about "T" header construction. This is fine as long as the header is relatively short, say 6' or less, and the ends of the header are nailed well to the king studs so that the header can't twist. The danger here is the top of the "T" (compression portion) buckling sideways under heavy load. This can't happen if the above two details are adhered to, but could if not.

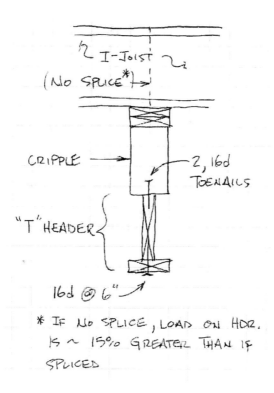

I-Joist
(No SPLICE*)

CRIPPLE → 2, 16d TOENAILS

"T" HEADER

16d @ 6"

* IF No SPLICE, LOAD ON HDR. IS ~ 15% GREATER THAN IF SPLICED

You might be wondering whether the load is truly uniform? After all we're only showing one cripple applying load over the length of the header – how is that uniform? Well, it's not actually. But as long as the distance between cripples is 24" or less, we can assume, for simplicity's sake, a uniform load over the length of the member. You could check the validity of this simplification by analyzing the header with a single point load. ProBeam™ makes that easy. You would find, however, that the Acceptable Solutions come out approximately the same. Try it.

Trimmer. Here is the required bearing length and applied gravity load at each trimmer.

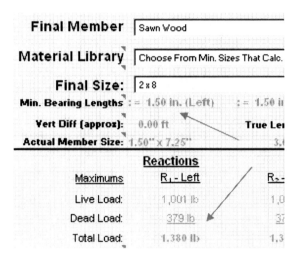

Final Member	Sawn Wood
Material Library	Choose From Min. Sizes That Calc.
Final Size:	2 x 8

Min. Bearing Lengths : = 1.50 in. (Left)	: = 1.50 in
Vert Diff (approx): 0.00 ft	True Le
Actual Member Size: 1.50" x 7.25"	3.(

Reactions

Maximums	R₁ - Left	R₂-
Live Load:	1,001 lb	1,0
Dead Load:	379 lb	37
Total Load:	1,380 lb	1,3

The required bearing length will allow a 2x4 trimmer so the only question left is whether that trimmer can take 1,380 lbs. We make that analysis just like previously using Column, Post, Stud Calculator™ and find that a 7' long, 2x4 Doug Fir No 2 makes it by a mile, 82%.

King Stud. Do we even need to calc the king stud? What load does it take? Well, being located indoors, out of the wind, there will be no significant out-of-plane lateral loading. Out-of-plane seismic loads on stick framed walls are negligible. If our trimmer is taking all the gravity load, what's left for the king stud? Not much – its main jobs are to secure the door frame and provide resistance to roughhousing when the kids play Nerf football. The king stud is okay by inspection.

Chapter 5
Floor Framing

A Quick Recap

To summarize what we've learned so far:

o *Green design conforms to building code but doesn't necessarily exceed it.*

o *More is usually not better.*

o *Beams, rafters, and joists are at their most efficient when: 1) Span is minimized; and 2) Member height is maximized.*

o *Use software to tell you how efficient a member is. Select options with low percentage of overdesign.*

o *Sometimes the greenest option is the one laying around in your boneyard.*

o *Engineers are not motivated by built-green. Their motivation, covering their own backside, is usually anti-green.*

Now let's examine floor framing.

Floor Joists

Designing floor joists is a lot like designing rafters, the process and variables being largely the same. There is one big difference, however. With floor joists we care A LOT about deflection. Recall when we designed roof rafters, deflection criteria were based on the ceiling material attached to the bottom side, drywall usually. We did not want rafters to sag so much as to crack the drywall.

Deflection criteria for floors is more stringent. Not because we're worried about cracked floor tiles or some other hard surface, but because we don't want the floor to feel bouncy. A stiff floor is one that doesn't deflect much under live load - people walking, dancing, dining, etc. Occupants like stiff floors, not bouncy ones.

Let's jump into an example.

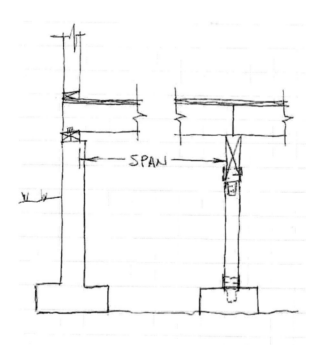

Span. We're not given a span so let's assume one, say, 14'.

Spacing. Joists are typically spaced at 16", a dimension driven by the plywood or OSB sheathing on top. Here is a green opportunity. What if we used 24-inch spacing? Or even 32"? We'd use fewer joists and more insulation. As an example, going back to our 40'x50', 2,000 square foot house, assuming joists span the 40' dimension, using 16" spacing, we'd use approximately 37 joists. Using 24" spacing, we'd use 25 joists, a savings of 12 joists, 480 lineal feet of floor joist material. *Significant*. Using 32" spacing we'd only use 19 joists. The tradeoff is that plywood subfloor thickness might increase.

How much? Let's check www.apawood.org, Engineered Wood Construction Guide – Roof Construction, Table 29, and find out. For APA Rated Sturd-I-Floor™, here are a few options:

o **5/8"**, 16 oc span rating, joists at 16", good for **100** psf live load.

o **5/8"**, 16 oc span rating, joists at 24", good for **40** psf live load.

o **3/4"**, 24 oc span rating, joists at 24", good for **100** psf live load.

o **3/4"**, 24 oc span rating, joists at 32", good for **50** psf live load.

o **7/8"**, 32 oc span rating, joists at 32", good for **100** psf live load.

o **7/8"**, 32 oc span rating, joists at 48", good for **40** psf live load.

These values assume dead load of 10 psf or less and deflection limits of l/180 for live + dead, and l/240 for live load only. Meaning if flooring is heavy tile or concrete, Table

29 doesn't apply; thicker plywood will be required. Also, this deflection criteria may be too flexible for some people's comfort, again, necessitating thicker material.

Back to our example. Let's assume flooring is carpet or vinyl (less than 10 psf dead load) and that we're okay with Table 29's deflection criteria. We see that for residential applications with live load of 40 psf, we can space our joists at 24" and use the same 5/8" plywood as if we spaced them at 16". Thus it costs us nothing extra to save those 12 floor joists. *Very green*. Let's continue using 24" joist spacing.

Loading. Live load for residential floors is 40 psf. Dead load is the sum of the flooring material, subfloor, insulation, plumbing and mechanical, and the joists themselves. I typically use 15 psf.

Depth For Insulation or Architectural Purposes. Many times the depth of a joist is not controlled by strength but instead by ceiling height or some other an architectural requirement. Let's assume that our example has no such constraint; we can use any depth of joist that's efficient.

Deflection Criteria. From Table 1604.3 in the 2006 IBC, maximum deflection for floors is:

o   Live only: l/360

o   Dead + Live: l/240.

I generally use more stringent criteria because I do not want a callback for bounciness. Here is the criteria I use for floor joists:

o   Live only: l/600

o   Dead + Live: l/480.

Since it's only a couple clicks on the computer, let's check both.

Calc It. Here's the input for our joist at 2-foot spacing and code minimum deflection.

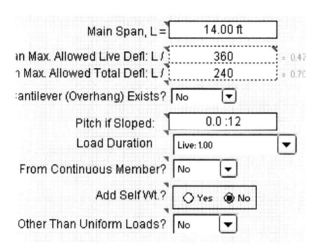

Remember with ProBeam™ each red triangle conceals a help note explaining that particular input. Here's our loading.

r Full Length of Member	Live, psf	Dead, psf	Tributary Width, ft
not including snow)			
Roof Snow (only)			0.00 ft
Floor 3 Loads			
Floor 2 Loads			
Floor Loads	40 psf	15 psf	2.00 ft
Wall Dead Load			

Done. If we wanted to use sawn material, here are our choices.

4x And Smaller (Lumber)

| Lumber Material | Douglas Fir-Larch | |
| Lumber Grade | No. 2 | |

Repetitive
Member Use?

Yes ▼

properties for what

Acceptable Solutions

2 x 12	(4) 2 x 8
(2) 2 x 8	3 x 10
(3) 2 x 8	4 x 8

You'll note that I took repetitive member credit because joists are spaced 24" or less and thus share load.

If we wanted to use 2x material, we'd select the 2x12 and be done. However, I prefer I-joists, so let's look at those options.

I-Level, TJI

11-7/8" TJI / L65	11-7/8" TJI 110
16" TJI / L65	9-1/2" TJI 210
16" TJI / L90	9-1/2" TJI 230
14" TJI H90	11-7/8" TJI 360

The 11-7/8" TJI 110 and 9-1/2" TJI 210 are the least expensive options. Either would be an acceptable choice though I would definitely select the 11-7/8 if it were my call. Here's why.

Remember, I don't want a callback for a bouncy floor. If we look at the results section for each of the above here's what we find. First we select the 9-1/2" TJI 210, like so.

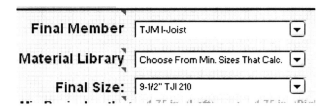

And here are the results.

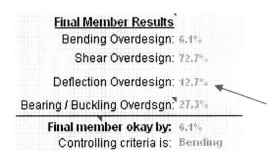

We see that this joist makes it by 6.1%, with the controlling criteria being bending stress. Very efficient. Looking further, however, we see that the Deflection Overdesign is only 12.7%. And that's using code minimum, bare bones deflection criteria.

So, while this joist meets code and is efficient, it is also likely to be somewhat bouncy.

Now let's select the 11-7/8".

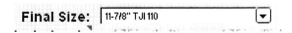

... and check its results.

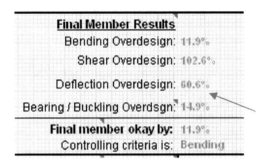

Final Member Results
Bending Overdesign: 11.9%
Shear Overdesign: 102.6%
Deflection Overdesign: 60.6%
Bearing / Buckling Overdsgn: 14.9%
Final member okay by: 11.9%
Controlling criteria is: Bending

This model is not quite as efficient – it makes it by about 12%. However, Deflection Overdesign is huge, 60.6%. This means that this joist is much stiffer than the 9-1/2" alternative and thus will yield a much less bouncy floor.

Now let's make another run, pumping up our deflection criteria.

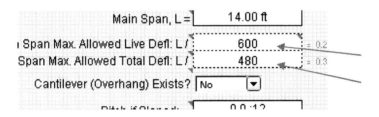

Main Span, L = 14.00 ft
Span Max. Allowed Live Defl: L / 600 = 0.2
Span Max. Allowed Total Defl: L / 480 = 0.3
Cantilever (Overhang) Exists? No
Pitch if Sloped: 0.0 :12

Changing the above two inputs is all that's required. Scrolling down to Results we find that the 2x12 is still okay. *What, how?* The reason is that the controlling criteria for 2x12 in both cases is bending stress, not deflection. In other words, a 2x12, while slightly weaker than the I-joists, is *stiffer*. This is the reason that in the early days of I-joists, when they were touted for their superior strength, an I-joist could seemingly span greater distances than similar depth 2x material. But they paid the piper in terms of bounciness. Sure I-joists might have been a little stronger, but they were less stiff, i.e. quite a bit more prone to deflection, and so got a reputation for producing bouncy floors. The I-joist industry has since learned their lesson and now uses l/600 deflection criteria in their designs.

Back to our example. Deflection controls the I-joists' design. Here're those results.

I-Level, TJI		
11-7/8" TJI / L65	14" TJI 110
16" TJI / L65	11-7/8" TJI 210
16" TJI / L90	11-7/8" TJI 230
14" TJI H90	11-7/8" TJI 360

Now there are no 9-1/2" options, and in the 110 series (the least expensive series) the minimum allowable height is 14". If we wanted to stay with 11-7/8" we'd need a 210 series.

Suddenly that 2x12 Doug Fir No. 2 alternative is looking pretty good.

Let's try one more tweak. I wonder how far off the 11-7/8" TJI 110 is?

ProBeam™ allows us to check any member whether it calcs or not via this dropdown.

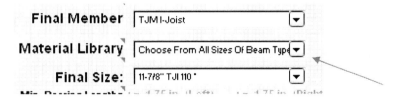

Now we can choose any member in that series' library, including the 11-7/8 TJI 110.

Here are the results.

```
Final Member Results
     Bending Overdesign:  11.9%
       Shear Overdesign:  102.6%
  Deflection Overdesign:  -3.6%
Bearing / Buckling Overdsgn:  14.9%
─────────────────────────────────
     Final member FAILS by:  3.6%
    Controlling criteria is:  Deflection
```

We see that in all of the *strength* categories - bending, shear, and bearing / buckling - this member makes it. Only in *deflection* does it fail, and only by 3.6%. Recall that we used an artificially high deflection criteria, selected somewhat at random. Unless the owners of this home were exceptionally insistent on a rock-hard floor, I would not hesitate to use this joist – it will still be much stiffer than code minimum.

Before we move on, did you notice that our joist was assumed to have a joint at the beam support? In other words each joist was not *continuous* over that beam, it was *simply supported*. When sizing joists, this is a conservative assumption. Let's say the lumber yard delivers I-joists 30' long and there winds up being no joint at the beam support. Is our analysis still valid? Yes, and it will be a little conservative. The floor will be stiffer than our analysis shows. When I design floor joists, I always assume a joint over the beam, mostly because I'm fearful that a framer will cut and hack his joists to make them easier to pack around, or to accommodate a utility, or to use up some old shorter material from another job, or whatever. But if he doesn't, I've been a little conservative, which, while not particularly green, yields a better floor.

Be careful, though, of floors designed by I-Joist manufacturers. Usually they've shaved their designs to the bone so that their price is as competitive as possible. Probably they are *depending on* the added stiffness of continuity over supports. If you come along cutting and hacking, adding joints, you're also adding bounciness and additional bending stress, and will likely void their warranty. Bottom line with manufacturer-designed I-joist systems: Install <u>exactly</u> per the manufacturer's plans and recommendations.

Floor Beams

Designing a floor beam is just like designing a header or any other beam. The variables are the same: span, deflection criteria, allowable depth, and loading. The real art comes in selecting the most efficient beam. Not so much because it's a single beam but because floor systems consist of many beams, posts, piers, and joists. Picking the wrong one has ripple effects throughout the system that can rack up inefficiency in a hurry. Here's an example.

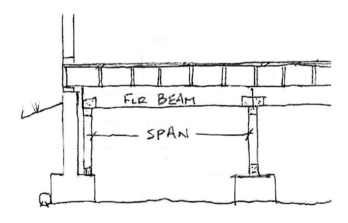

Span. We're not given a span so let's assume one, say, 8'. As we move through this example we may come back and tweak this span if advantageous to greening our design.

Loading. Live load for residential floors is 40 psf. Dead load is the same as the floor joist example, 15 psf, plus the self-weight of the beam which we add via radio button in ProBeam's™ General Inputs section.

Depth For Clearance or Architectural Purposes. If we're working with a crawl space of limited height we may want a shallow beam. For this example, however, let's assume that is not the case, depth isn't a concern.

Deflection Criteria. In our floor joist example we were quite concerned about deflection. With floor beams, however, it's not as critical because usually we're not dealing with long spans. Bounciness is especially troublesome when spans exceed about 14'. Our floor beam will span much less than that, probably 10' at the most so we can use code-minimum deflections, l/360 for live load only and l/240 for total load.

Tributary Width. This is half the distance to the next beam's centerline on one side plus half the distance to the centerline of the beam on the other side. To coincide with the joist span we set in the previous example, and assuming our floor beams might wind up being 6x (a conservative assumption), tributary width is, 14.5'/2 + 14.5'/2 = 14.5'. Note that if we wanted to use a 4x floor beam, the tributary width would be 14.3'.

Load From Continuous Member? In the General Inputs there is a Yes-No dropdown asking this question. If there's a possibility that floor joists will be continuous over this beam we select Yes. If we're 100% sure that all the floor joists will be spliced over this beam we can select No. Selecting Yes is always conservative and adds 15% to our uniform loads. I generally select Yes when designing floor beams.

Calc It. Here's the general input.

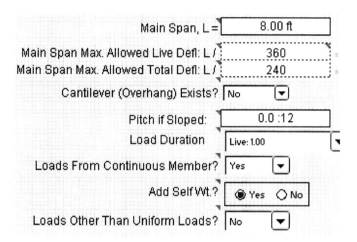

... and here's the loading.

II Length of Member	Live, psf	Dead, psf	Tributary Width, ft
:luding snow)			
of Snow (only)			0.00 ft
Floor 3 Loads			
Floor 2 Loads			
Floor Loads	40 psf	15 psf	14.50 ft
all Dead Load			

Checking results for Doug Fir No. 2, we see that the smallest sizes are 4x14 and 6x12. Those are a little big and expensive – probably not a good choice. Looking further down the list of acceptable alternatives, however, we see some attractive engineered lumber options, for example a 1-3/4 x 11-7/8, 1.55E LSL.

Here's where it gets interesting.

Even though we could use that relatively inexpensive LSL we wouldn't. Why? Mostly because it's too narrow to provide bearing area for the joists. If there happened to be two joists butt-spliced over this beam, each would require at least 1.5" of bearing length, or 3" total. Clearly a 1-3/4" wide beam doesn't cut it. So if any joists are spliced over this beam we're stuck with at least a 4x sawn or 3x engineered member.

Something else to consider is the load on our pad footing.

	Reactions	
Maximums	R_1 - Left	R_2 - Right
Live Load:	2,668 lb	2,668 lb
Dead Load:	1,048 lb	1,048 lb
Total Load:	3,716 lb	3,716 lb

These are the reactions from the analysis we just performed. Each end of our beam brings about 3,700 lbs to its footing. Keep in mind that each pad footing supports TWO beam ends, or ~ 7,400 lbs. That can be a lot depending on the quality of our soils. Average soils can support about 1,500 psf (pounds per square foot). Here is a simple equation to calculate the approximate required footprint of a square footing:

$$[(rxn1 + rxn2) / asb]^{1/2} = B$$

Where rxn1 = reaction from one beam end on footing, lbs.

rxn2 = reaction from the other beam end on footing, lbs.

asb = allowable soil bearing pressure, psf

$^{1/2}$ = quantity in [braces] raised to the 1/2 power, which is another way of saying square root of the quantity in [braces].

B = length of a side of the square footing, ft.

For our case, $[(3716+3716) / 1500]^{1/2} = 2.2$'. That's a pretty big footing – most residential pads are no more than 2' square or round.

So here's the upshot. Our design as it stands is forcing a floor beam that's big and expensive, and we're also forced to use a pad footing that's unusually large.

Here are a couple alternatives:

1. Reduce the span of the floor joists. This could easily result in smaller, less costly joists. It will also lessen the tributary width on the beam and footing, making each smaller and more economical. The tradeoff is an additional beam, post and pier line.

2. Reduce the span of the floor beam. Moving the pad footings closer together will result in smaller beams and smaller footings. The tradeoff is that additional footing(s) will be required.

Thus, laying out an efficient floor system is a balancing act that's usually accomplished by trial and error. Fixing a variable or two, however, can simplify things.

Let's say, for example, we want to use 4x10 Doug Fir No 2 for floor beams. And also we're happy with the span of our floor joists. The challenge, then, is to determine how far we can span those 4x10s.

ProBeam™ provides the answer. The strategy is to adjust the span until the 4x10 just calcs. Let's do it. First we adjust our tributary width to accommodate a 4x floor beam (as opposed to a 6x). It drops from 14.5' to 14.3' – not a lot but every little bit helps. Right now our span is 8'. Let's try 7'.

Making those two changes gets us to a 4x12, still too big. Dropping the span a few inches at a time and recalculating, I wind up with 6.25', which results in our 4x10 making it by 0.8%. Very efficient.

The reaction is now 2,852 lb. per beam end, requiring a square footing, $[(2852+2852) / 1500]^{1/2} = 1.95'$ per side. Perfect.

A note of caution. The above analysis assumes uniform load on the beam, i.e. no point loads from posts or bearing walls above. When those conditions occur, and they will, they must be accounted for in ProBeam™, which is pretty simple, actually, it's just the remembering-to-do-it part that sometimes causes problems.

Posts and Piers

Posts. Posts in a crawl space are typically 4x4 material. Is that the right choice? Could we use something less expensive? Do we need something stronger? Before we get to the answers, let's explore exactly what those posts do, and what they don't do.

Typical posts in a crawlspace resist gravity loads from floor beams. They do NOT take lateral loads unless specifically designed and detailed for that purpose, which is so rare I've never seen it done. Why?

The reason that crawl space posts do not take lateral loads is because the lateral loads from the house above are transferred to the *perimeter footings* through shear walls and floor diaphragms. Perimeter footings, consisting of bolted mud sills, stem walls, and continuous footing strips are capable of transferring in-plane shear from above to the soil below. Small isolated pad footings connected to 4x posts with flimsy framing hardware are not capable of resisting any significant lateral loads.

Given that, is there ever a need for big, herking bolted connections on pier posts? No, unless specifically designed and detailed that way by an engineer who is trying to accomplish something unusual. Instead, save money by using cheap framing hardware, such as PB and PC series by Simpson Strong-Tie™. Or you could use 1x cleats. Or even toenails will work – anything strong enough to hold the post in place should a worker bump into it.

Some builders like to connect their posts after the piers are poured and cured. That's fine. You could use a simple angle clip, such as A34 by Simpson™ and powder-actuated nails, a.k.a. "Power Pins."

Now let's check the capacity of a 4x4 Doug Fir No. 2 post. As we learned previously, the capacity of any post, column, or stud depends a lot on its unbraced length. The unbraced length of a pier post is its full length.

For our example, let's assume we have a 3' tall crawl space, like so.

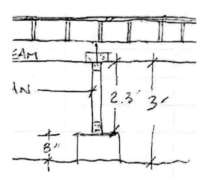

For our loading, let's use the loads from our floor beam example. Here are those reactions.

Reactions

Maximums	R₁ - Left	R₂ - Right
Live Load:	2,056 lb	2,056 lb
Dead Load:	797 lb	797 lb
Total Load:	2,852 lb	2,852 lb

Remember, there are two beam ends on our post, so the individual reactions listed above must be doubled.

Here is the general input from Column, Post, Stud Calculator™.

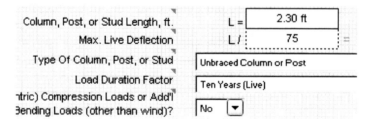

... and here is the loading.

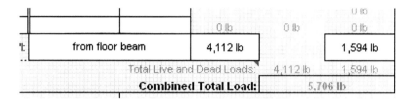

There is no wind load in a crawl space so we leave that section blank. Also, did you notice that we relaxed the Maximum Live Deflection to l/75? Since this post isn't connected to any cladding of any sort, we don't really care if it deflects (bows) much. By doing this we may get by with a smaller, less expensive post, but one that still meets all code *strength* requirements.

Here are our results.

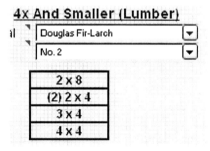

A 4x4 makes it and so does a double 2x4. Either would be okay. We wouldn't select the 2x8 for a couple reasons. First, it's no greener than the other choices, and second, it would not provide enough bearing area for the beams.

So to summarize, a 4x4 pier post is a very good choice.

There is one thing we would could do a little greener, though. Use a lesser quality, cheaper 4x4 - for example, Spruce Pine Fir.

Making that selection...

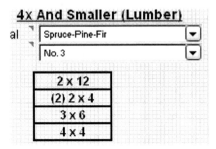

...yields essentially the same sizes as before. *Yes* – here is a nice little savings that, added to many others we've made, becomes significant.

<u>Piers, a.k.a. Pad Footings or Pads.</u>

When I look at a foundation plan, my primary goal is to take out as many unnecessary pad footings as possible. Here's how.

Pads near perimeter footings. If a floor beam terminates at a perimeter footing, why not use it rather than constructing a separate pad there? The perimeter footing, if in conformance with building code, will have plenty of capacity for a few floor beams.

There are several ways of supporting a floor beam on a continuous footing. The following option, though limited, illustrates my favorite (courtesy Simpson Strong-Tie™). The reason I like this so much is that no post is required at all.

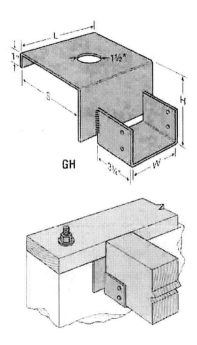

The limitations are as follows.

- o It only accommodates four girder sizes - 4x6, 4x8, 6x6, and 6x8. Too bad, I often like using 4x10.

- o The maximum allowable load is 2,000 lb. In our previous example, the load was 2,852, thus even if the beam would have fit in the hanger, the load would have been too great.

Nevertheless, where this works, it's very green.

In the case where floor joists are hung from a girder, the girder should be extended onto the mud sill for bearing, like so.

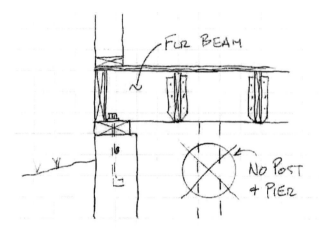

In the case where joists bear on top of the girder, bear the beam on a post on the flange of the footing.

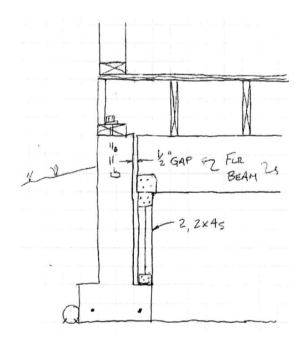

It this detail, it's important that a horizontal rebar exists in the footing below our post to help distribute the post's load along a generous length of footing.

Pad Under Point Loads. Some designers think that wherever a post bears on a floor, a footing is required directly under it. Sometimes this is true but many times the point load can be picked up by a beam or double joist and taken to another footing nearby.

Using Three Pads Where Two Will Do. Say you have an interior wall 15' long bearing on the first floor. Do you install three posts and piers under it 7.5' apart? Do that and you'll waste a post and pier. Why not use two, and cantilever the ends of the floor beam? Like this.

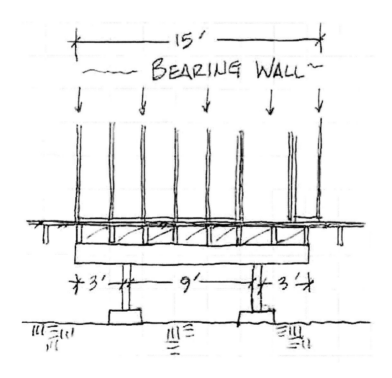

Let's calc it. First the beam. To accurately calc the beam requires software that can handle cantilevers on both ends. ProBeam™ does not have that capability, however, ProBeam™ can handle a cantilever on one end, which will yield approximately the same results, though slightly conservative (2% to 9% overdesign - always safe). Here is the general input.

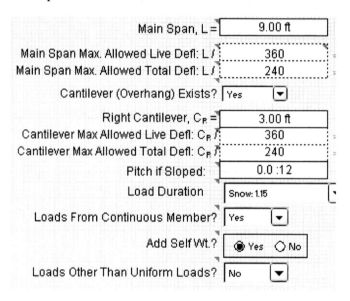

Now Loads. Let's assume our bearing wall applies loads from roof trusses that bear on it. Say the trusses span 30' and our wall is the only interior bearing. Tributary width, thus, is 30'/2 = 15'.

We also have dead weight from the wall itself, which for interior walls is normally 7 psf. The tributary width is the wall's height, say 8'.

For floor load, let's assume our bearing wall occurs between post and pier lines that are 14' apart. Tributary width on our floor beam is thus 14'/2 = 7'. As usual, we assume the floor joists are not spliced over our beam, so in the General Input section we select "Yes" to "Loads From Continuous Member?".

Here is the loading input.

Full Length of Member	Live, psf	Dead, psf	Tributary Width, ft
ıt including snow)		15 psf	15.00 ft
Roof Snow (only)	25 psf		15.00 ft
Floor 3 Loads			
Floor 2 Loads			
Floor Loads	40 psf	15 psf	7.00 ft
Wall Dead Load		7 psf	8.00 ft

Done. The smallest Doug Fir No 2s that make it are 6x14 or 4x16 – too big. Here are the glu-lams that calc.

Glued Laminated Members

e 24F-V8 (DF/DF) ▼

Acceptable Solutions

2.5" x 13.5"	5.125" x 10.5"
3" x 12"	6.75" x 9"
3.125" x 12"	8.75" x 9"
5" x 10.5"	

Note the grade I selected is 24F-V8, which is required for cantilever and multi-span cases.

Here are the PSL solutions.

2.0E Parallam PSL	
-	5-1/4" x 9-1/4"
2-11/16" x 11-7/8"	7" x 9-1/4"
3-1/2" x 11-1/4"	Slf Wt= 35

I like the PSL option best because it's the most efficient. Remember, though, if joists are butt-spliced over this beam, we need a 3" wide member for bearing area, the 3x12 glu-lam most likely. We're assuming no joist splices so we pick the PSL.

We'll need the reactions to size our post and footing. Here they are.

	Reactions	
Maximums	R_1 - Left	R_2 - Right
Live Load:	3,390 lb	6,026 lb
Dead Load:	1,816 lb	3,631 lb
Total Load:	5,205 lb	9,657 lb

We're only interested in the heavier reaction because our beam actually has cantilevers on both ends, not just one as in this simplified analysis.

Looking at the Final Results, we see this red warning note.

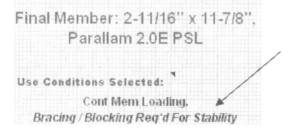

Final Member: 2-11/16" x 11-7/8", Parallam 2.0E PSL

Use Conditions Selected:
Cont Mem Loading,
Bracing / Blocking Req'd For Stability

What does that mean? With any cantilever there is a negative moment at the support. Negative moments in beams cause the bottom of the beam (compression side) to twist sideways. If there is nothing to prevent that outward movement, the beam can become unstable and flop over. To ensure this doesn't happen we need to provide a lateral brace. Here is how I'd do it using a scrap piece of 2x4.

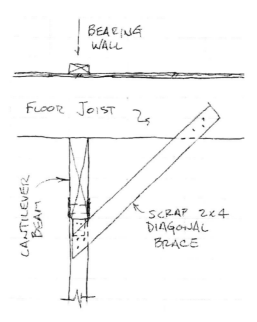

I checked the post and found that 2, 2x4s work. Excellent.

We could determine the footprint of our footing using the simple equation presented earlier in this chapter but instead let's design the whole thing, rebar and all. For this we'll use FootingCalc™.

FootingCalc™ is a very powerful program with quite a bit more capability than we need for this simple footing. Thus several input sections are not used and remain hidden. Here is the General Input.

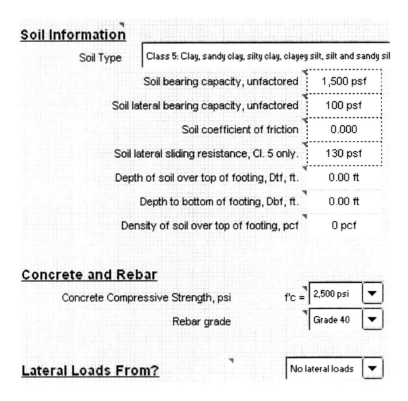

Our loads come from ProBeam™ and are entered similarly as we did when sizing a post, like so.

from beam analys	6,026 lb		3,631 lb
gravity loads, unfactored:	6,026 lb		3,631 lb

Now we can get down to the business of sizing the footing. This is a trial and error process, but software makes it easy.

With just a few trials I come up with the following.

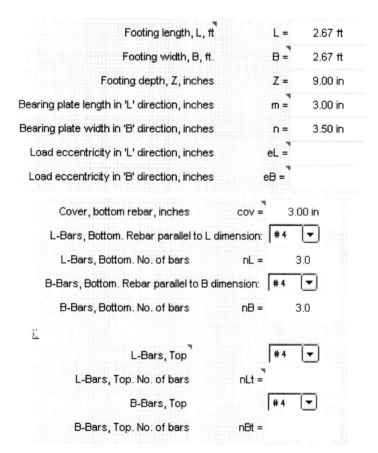

Footing length, L, ft	L =	2.67 ft
Footing width, B, ft.	B =	2.67 ft
Footing depth, Z, inches	Z =	9.00 in
Bearing plate length in 'L' direction, inches	m =	3.00 in
Bearing plate width in 'B' direction, inches	n =	3.50 in
Load eccentricity in 'L' direction, inches	eL =	
Load eccentricity in 'B' direction, inches	eB =	
Cover, bottom rebar, inches	cov =	3.00 in
L-Bars, Bottom. Rebar parallel to L dimension:	# 4	
L-Bars, Bottom. No. of bars	nL =	3.0
B-Bars, Bottom. Rebar parallel to B dimension:	# 4	
B-Bars, Bottom. No. of bars	nB =	3.0
L-Bars, Top	# 4	
L-Bars, Top. No. of bars	nLt =	
B-Bars, Top	# 4	
B-Bars, Top. No. of bars	nBt =	

I didn't use any rebar in the top because there is no moment or uplift on our footing. So 3, #4 each way in the bottom works fine.

The final design is summarized at the bottom of the sheet.

	Final Design
Conc f'c:	Footing Size: LxBxZ = 2.67-ft x 2.67-ft x 9-in
2500 psi	L-Direction Rebar: 3, # 4 bottom + 0, # 4 top
Rebar:	B-Direction Rebar: 3, # 4 bottom + 0, # 4 top
Grade 40	No Plinth
Soil Brng Cap.	
1500 psf	No Hairpin
Inclined Load:	
0 deg.	No Anchor Bolts
(no incl'n)	

This might seem like a large footing but keep in mind that it's supporting a big chuck of roof as well as a generous portion of floor. And, our soil, class 5, is not so terrific.

What if instead of our two-footing, cantilever beam we used three footings spaced 7.5' apart and a simply-supported beam? With just a few input changes here's what I come up with:

* Beam – 4x14 or 6x12 Doug Fir No 2, or 3x9 glu-lam, or 2-11/16x9-1/4 PSL.

* Post – same as previous.

* Footings – Middle footing is the same as we just designed, but the outer two footings can be smaller, 1.8' square.

So the beam could be a little smaller, though still rather large, and we'd use three footings instead of two, though the outer two would be a little smaller than in the cantilevered case. If I'm trying to build green, I'd go for the two footing, cantilevered beam option.

Summary and Wrap-Up

In these pages we've examined some of the more common applications of green framing. Certainly, there are others we didn't cover. But what is green framing, really? To me it's knowledge - an understanding of how beams, columns, shear walls, and footings work - and then applying that knowledge *efficiently* to any aspect of framing. A beam is a beam whether it occurs in a roof, wall, or crawl space. Learn how to design one and you can design one hundred.

There is a good reason that inefficient, wasteful framing is so rampant today: Builders and designers haven't been taught structural concepts. In years past, prior to computers and good software, this was, perhaps, excusable. Because even if the knowledge was there, extending it confidently to code-compliant, efficient design was so cumbersome it was nearly impossible.

Fortunately, those days are behind us. There's no excuse for building "old school." The claim, *"I've been doin' it that-a-way fer 50 years and never had one fall down yet,"* may be true enough, but it sure doesn't produce a green structure. Nor likely a code-compliant one.

Inefficient framing costs big – thousands of dollars per new home – *billions* worldwide annually. And *millions of tons* of timber, concrete, and steel are literally wasted every year.

What if each framer and designer understood the principles of green framing? What if engineers sharpened their pencils on every project? What a contribution the building industry could make to the green movement, to our planet. Talk about win-win!

CPSIA information can be obtained at www.ICGtesting.com
Printed in the USA
LVOW10s1946120615

442295LV00019B/211/P